= .= BEALE TREASURE =. =
= MAP TO =
= = CIPHER SUCCESS = =

– STAN CLAYTON –

An environmentally friendly book printed and bound in England by
www.printondemand-worldwide.com

PEFC Certified

This product is from sustainably managed forests and controlled sources

www.pefc.org

This book is made entirely of chain-of-custody materials

www.fast-print.net/store.php

Beale Treasure Cipher Success

First English Edition published 2012.

ISBN 978-178035-347-0

First published 2012 by
FASTPRINT PUBLISHING
Peterborough, England.

Foreword

Come on you computer whiz kids the race is now on, win fame and fortune by being the one to solve the famous Beale Treasure Location cipher. I started on the codes in 1965 and got this far using pen and paper and I've written this book in the hope that someone will write a programme and use my methods to find the answers to the rest of it.

Many other bits that didn't make sense I left out like TWO TON KIT TEAM, and VALUED AT, because nothing else fitted in with it. I also have PLOT and two initials after it coupled with a long name of eight and four letters. I got it so you can get it as well, however I haven't got the name of the cemetery.

The name is hidden in the story in two separate parts in the hope that someone finds it so I don't have the responsibility of cemetery's being desecrated. I know readers will believe I'm just another liar saying I have it without showing proof, but I'm 79 and could pop my clogs any time so I have nothing to gain by lying.

If no one gets it I will publish the other half of my Space Saga and promise to put it in the story. Sample:

501 - 823 - 216 - 280 - 34 - 24 - 150 - 1000 - 162 - 286 - 19 - 21 - 17 - 340 - 19 - 242 - 31 - 86 - 234 - 140 - 607 - 115 - 33 - 191 - 67 - 104 - 86 - 52 - 88 - 16 - 80 - 121 - 67 - 95 - 122 - 216 - 548 - 96 - 11 - 201 - 77 - 364 - 218 - 65 - 667 - 890 - 236 - 154 - 211 - 10 - 98 - 34 - 119 - 56 - 216 - 119 - 71 - 218 - 1164 - 1496 - 1817 - 51 - 39 - 210 - 36 - 3 - 19 - 540 - 232 - 22 - 141 - 617 - 84 - 290 - 80 - 46 - 207 - 411 - 150 - 29 - 38 - 46 - 172 - 85 - 194 - 39 - 261 - 543 - 897 - 624 - 18 - 212 - 416 - 127 - 931 - 19 - 4 - 63 - 96 - 12 - 101 - 418 - 16 - 140 - 230 - 460 - 538 - 19 - 27 - 88 - 612 - 1431 - 90 - 716 - 275 - 74 - 83 - 11 - 426 - 89 - 72 - 84 - 1300 - 1706 - 814 - 221 - 132 - 40 – 102 - 34 - 868 - 975 - 1101 - 84 - 16 - 79 - 23 - 16 - 81 - 122 - 324 - 403 - 912 - 227 - 936 - 447 - 55 - 86 - 34 - 43 - 212 - 107 - 96 - 314 - 264 - 1065 - 323 - 428 - 601 - 203 - 124 - 95 - 216 - 814 - 2906 - 654 - 820 - 2 - 301 - 112 - 176 - 213 - 71 - 87 - 96 - 202 - 35 - 10 - 2 - 41 - 17 - 84 - 221 - 736 - 826 - 214 - 11 - 60 - 760.

221= 5 + 12 = 17 + 11 = 28 = B
736=16 + 5 = 21 + 6 = 27 = A
826=16 +16 = 32 + 6 = 38 = L
214 = 7 +16 = 23 + 6 = 29 = C
11 = 2 + 7 = 9 + 6 = 15 = O
60 = 6 + 2 = 8 + 6 = 14 = N
760 =13 + 6 = 19 + 6 = 25 = Y

To encode A = 27 minus 6 = 21 minus 5(the sum of the previous code number) which leaves 16 so any number that adds up to 16 could serve as code number i.e. 88,97,1906,484,925,592,736,

I believe the added number 6 is part of the letter from ST LOUIS.

"S" can be 19 or 7."T" is 20 or 6. "L" is 12 or 14 "O" is 15 or 11 "U" is 21 or 5, "I" is 9 or 17.

The letters of St Louis are used as plus or minus add ons all through the code, it appears the letter we've all been searching for has been with us all the time.

1014 lbs GOLD
3812 lbs SILVER
DEPOSITED NOV 1819
1907 lbs GOLD
1288 lbs SILVER
DEPOSITED DEC 1821

= CHAPTER ONE =

Bloody Hell. What hit me, my pain racked brain failed to catch the memories flooding through the inky blackness of a returning consciousness. Have I had an accident, been beaten up, or what, my mind was in turmoil trying to restore order.

I can't see, anything I'm blind, panic set in. My guts knotted into spaghetti, as I got more warnings, I can't move, I'm paralysed, my arms and legs won't work, wriggling my fingers and toes brought a moments relief, instantly replaced by one of utter dread, I've been buried alive, I'm firmly jammed in a coffin, I can't bend my knees and bending my heads no use "Oh Christ, help me" I sobbed realising the hopelessness of the situation.

There's not a sound to be heard I must be buried deep, trying to hold my breath to suffocate and get it over with, only ended up with me gasping for air.

Preoccupied with fear, I hadn't noticed the blackness turn to a light grey, then a ray of light shone through, lifting my head a little I saw a small figure limping towards me. My prayers have been answered, and he's coming, the intense bright halo of light streaming from behind him put his face in shadow, blinking and shaking my head to dislodge the tears from my eyes I lay in awe.

What can I say at this once in a lifetime occasion, coming closer I saw the reason for his limp, a gun barrel tied to his leg for a splint. This completely flipped me, My Saviour with a broken leg, how the hell could that be possible, was he trying to tell me something, my brain refused to function rationally, and I knew I was going mad when he bent over me and I saw through the open doorway behind him the most amazing sight, which cancelled out any reasonable thoughts I had.

Two chairs faced a large television screen with star maps and coloured tracks showing on it, I know I'm not dreaming, we don't dream in colour. My befuddled brain tried to unravel more of the mystery, My saviour in a space ship, the word Spaceship triggered my thought process into action. Now I'm beginning to remember, this little chap bending over me is the one I saved from the bull, and must be the pilot, I expected him to check if the straps tying me to the floor were tight, but he surprised me by undoing them simply by sliding a flat sleeve along them. Looking at them later, I reckon a kid of two could have undone them, they were just two overlapped serrated straps with a flat sleeve which pulled over the serrated parts to lock them together, to undo just move the sleeve to a non-serrated part, if this is an example of their technology I'm interested in getting all the new ideas I can.

Exhilaration tingled through my body as I realised I was thinking rationally again; The little chap helped me to my feet then held my arm to steady me, waiting until I could stand on my own, he dipped a beaker into a bucket attached to the wall filled it took a long drink, then handed it to me, I drained it in one go, like the nectar of the gods it cleared my head Now I began to remember with a rush, this must be water for our bull this chap was supposed to kidnap. thinking of Bill our prize Hereford bull brought more memories flooding back, as if it was glad to be in its rightful place.

This euphoria didn't last long as reality set in. What the hell have I let myself in for I'm a stowaway on a spaceship going at thousands or millions of miles an hour, to god knows where, I've no idea how long it will take, it could be years, and what happens at journeys end, will I be a prisoner, a slave or what. Will I ever return to Earth, and why is this little chap running around with my gun barrel tied to his leg, this part of my memory eluded me no matter how hard I tried, it just wouldn't reappear, although everything else is improving. Now my circulation has returned my pains disappeared, and I doubt if I was beaten up and kidnapped in the first place, I had another puzzle to sort out how could such a weedy looking chap manage to drag me in here. Looking around there was only a bit of hay in an old fashioned manger fastened to the wall next to the water trough, and the leather straps lying on the floor.

I'd expected to see stacks of food containers and piles of hay, but there was nothing, this lack of food can only mean our destination can't be far away. Which brings to mind our nearest neighbour the moon, scientists have been looking all over space for extraterrestrials when it looks like they are living on the far side of the Moon, if that's the case it wont be long before we find them, the way the Germans and the Americans have developed their rockets they will soon be exploring space. And I'll be there to welcome them, imagine their faces when I step forward to shake hands and they're dressed in large cumbersome space suits, looking more like deep sea divers. Now I have to remember how I came to be in this predicament in the first place, going back as far as I can, seems to be the best approach, then each returning memory will trigger others and bring me back to the present, it may be a good idea to think of some recent pleasant memories they tend to stick in your mind better than nasty ones.

The trouble is where do I start, school times seem a good place but I'll get bogged down remembering all the things that happened there, I think its better to start from when I left school and got my first job, which I can certainly remember because I got it for being lazy, our gang got a job on a local farm picking up cider apples, on their second day I joined them, they had to shake the apples down then pick them into buckets and tip them in a high sided trailer, this was back breaking so I got the lads to push the trailer under the trees and shook the apples straight into it, because we cleared the orchard in record time, he offered me a job.

= CHAPTER TWO =

Returning to work on the farm after spending three years National Service in the Air Force servicing the latest jet fighters and army planes was quite a change but I was enjoying myself, the balmy summer days and the quiet pace of life suited me, working a few days a week for pocket money and with over three hundred acres of woods and meadows to shoot in life couldn't be better Another beautiful sunny day found me sat on a straw bale in the old wooden barn eating my sandwiches, surrounded by a score of chickens busily scratching around and squabbling over the few crusts I occasionally threw them, just to keep their dusty scratching at a distance.

Life was so peaceful and quiet I nearly fell asleep; jumping up quickly to wake myself, I threw my leftovers to the fowls' to make up for alarming them, and set off to return my cup to the farmhouse, about half-way there a sudden explosion of terrified squawking and flapping wings erupting behind me stopped me in my tracks, dropping my cup on the garden wall I dashed back to the barn, rounding the corner I was too late, a pile of feathers lay on the ground and a tornado of whirling dust and feathers billowing over the battlefield, showed me Reynard one of our local foxes had outwitted me and collected his dinner for the day, in the short time I'd been away he had struck, biting the head off one chicken, and the trail of feathers showed he had taken another with him.

It made me mad to think the crafty devil had lain nearby, watching and waiting for me to move away then he had pounced, I felt he was laughing at me, getting his own back on me for shooting one of the pack a week ago.

Picking up the headless chicken, I took it up to the farm house and gave it to Cecil the old farmer, he knew what had happened by the noise of the commotion and wasn't too happy about losing more hens, he even talked about calling in the local hounds, which showed how serious the fox problem was getting, no farmer wants the hunt smashing down fences and upsetting the cattle, the thought of the hunt charging around the farm in full cry didn't appeal to me, I offered to go and hunt them and reminded him I'd shot one a week ago, but he said "don't bother yourself lad, you were lucky, if the mighty hunter Fred Nibblet couldn't get any, you don't stand a chance."

Fred the cowman from the next farm, was renowned for his poaching skills, that was until one of our foxes raided his fowl pens and killed all his chickens, he boasted in the Virgin Tavern that he'd soon have a row of fox heads lining the bar. When no heads appeared after a few weeks he became a laughing stock and stopped going there.

Listening to the old farmer referring to the foxes ability to outwit everyone made me think, they always seemed to know what we were doing, if we hid a gun on the tractor, even under a coat we saw no sign of them, but if we had no gun they would walk beside the tractor laughing at us, this gave me an idea, collecting my gun and a handful of bullets from the cowshed where I kept it, I shoved it through my belt jumped on my motor bike and raced off down the road to the foxes den in the lower fields. Riding along gave me time to think of a way of outwitting the sentries at the dens, half a mile from the warren of fox holes.

I dismounted, collected a pile of sticks tipped some petrol from the tank over it and lit it with my new fag lighter, then dropped a couple of bullets in the middle of the fire, I moved a hundred yards away and did the same again. Leaving the bike I wormed my way to the dens, I was nearly there when the first bullets exploded, and I actually saw two young foxes streak off to investigate. Now I had lessened the chance of being seen I moved faster, but I was nearly too late, Reynard came trotting along from a different direction to the one I expected, dragging his kill. He dropped it and lifted his head to listen as excited yapping came from behind me, the sentries had discovered my scent and were giving out a warning, knowing I wouldn't get a better chance I quickly lined up my sights making due allowance for the slight wind and squeezed the trigger, the smoke cleared and the fox was still standing motionless, as if listening to the noise of the bullets exploding in the second fire.

"Blast, I've missed," all my scheming had come to nothing, hastily ejecting the spent cartridge I quickly reloaded and swiftly took aim. Before I could fire the fox buckled at the knees and flopped to the ground, running over to him holding my gun at the ready just in case, I gazed down at the body slumped at my feet, I felt the exhilaration of hitting the target at such an extreme range turn into guilt, killing for the sake of killing put me into the same category as him, he had been a worthy opponent, but making an attack on the farm in broad daylight was really too much.

Using my bike and going the long way round to get in front was cheating, but all's fair in love and war, perhaps he had been distracted by the thought of the hero's welcome awaiting his return with his ill gotten gains, and now he'd paid the ultimate price for his lapse of attention.

A gust of wind ruffled the hair on the body giving the impression of life in it, I stepped back and used my rifle to turn the body over, at first I couldn't see any sign of injury, I even thought he may be playing dead then noticed with relief a small bullet hole beneath his ear. I know it may sound funny but I felt uneasy about shooting a dead body, especially as I was the one who shot him in the first place.

Throwing the body on the hedge and covering it with thorn hedge cuttings, to stop the crows pecking at it, in case the farmer wanted it for a glass case, I tied the dead chicken to the handlebars and returned the long way to the farm, The farmer had heard the noise of my bike and was waiting at the drive gates, he noticed the dead fowl and nodded his approval, "I see you got him," he said "he wouldn't have given that up unless you killed him, although it took a lot of bullets by the sound of it," I had to explain how I had distracted the other foxes, expecting praise for using my initiative I was disappointed when he said.

"You can't use that idea again, they won't get caught twice."

Later celebrating with the farmer over a cup of tea in the farmhouse, I thought it over, now was the time to hit them again, losing two from the pack would upset the hierarchy, while they were preoccupied and fighting among themselves I'd stand a better chance of bagging a few more, I've outwitted them once and can do it again, this time I have a more elaborate plan, which includes getting two nights food supplies and a few Crowscarers, these are fireworks tied to a piece of rope, when the rope is lit it smoulders setting of the fireworks tied along it at allotted time intervals.

Next morning being Friday I put my plan into action, loading up the tractor for a days muck spreading, I put my gun and supplies in a plastic bag hid them in the cab then drove down by the lower fields I knew the foxes would be watching me so I threw out lumps of meat to distract them while I set my Crowscarers to go off at different times of the day.

This was to make them stay in the den and have to come out to hunt tonight, last of all I hid my gun and supplies in a ditch, covering them up with leaves, and a few forkfuls of muck off the trailer, to disguise the smell of humans.

Finishing the last load of the day I returned to the farm put the tractor away, then called at the farmhouse to let him know what I was doing in case he called the police when he heard shooting in the night, He insisted I had tea with him and by the time we finished it was getting late, Setting off to begin my war against the foxes I used different tactics to confuse them, such as turning back on my tracks, then running like mad to pop up in a different place, eventually tiring of the game I sat on the butt of a large oak tree just inside Warndon Woods.

Hidden on three sides by bushes with a clear view of their burrows, my field craft had certainly paid off, I could tell by the short sharp yapping calls of the foxes they were confused and worried by my antics. Watching five foxes leave the den for their nights hunting was very interesting and I lined up my sights a few times, but by shooting one now the rest would go to earth for days, letting them out meant I could shoot each one when they returned with their nights kill.

Watching the night unfolding was unforgettable once the crows and Jays had finished their raucous calling the dramas had to be seen to be believed, I watched mesmerised as a large white Barn Owl swooped silent as a ghost over the tree I was under picked up a mouse or something and with superb flying skill went into an almost vertical climb to clear the next tree, after a while I dozed off, this wouldn't upset my plans as I wasn't expecting any activity until dawn.

Something woke me about two hours later and I sat wondering what had disturbed my sleep, settling down again listening to the noises of the night and recognising most of them a new sound caught my attention, a drumming noise seemed to come from the ground, jumping up to look over the bushes I expected to see animals running by, but the bright harvest moon showed nothing, then I heard a loud snort and knew instantly.

It was old Bill our prize Hereford bull in the next field, and something was upsetting him. Rustlers, the thought struck me. Newels prize Hereford bull had disappeared from a neighbours farm two year ago which caused a sensation at the time, but he was never found, now I was in a position to catch them red handed, grabbing my gun in one hand and bag in the other I raced in the direction of the commotion Old Bill was causing. by his antics.

Rounding the corner of the wood the sight that met my eyes didn't register at first, then I was horrified to see in the bright moon light our normally placid bull tossing something that looked like a body high in the air, when it landed he scooped it sideways and tossed it again, another body form on the ground nearer to me appeared to be crawling to get away from the enraged bull.

If the sight of the action taking place wasn't enough to amaze me the object he was crawling to was enough to stop me in my tracks, what I had taken to be a lorry with its tailboard down and light inside, was much too large besides it was dome shaped, and by the way it shone in the moonlight it was metal. This reminded me, we had been puzzled by the appearance of deep furrows in this corner of the field about a month ago. We thought it may have been an aeroplane trying to land, a Tiger Moth training plane had landed in this very field when it ran out of fuel at the beginning of the war, so it was assumed it had happened again, even though the aerodrome had been abandoned at the end of the war.

To this day I don't know why I did it, as I ran past I threw my gun and bag inside, then ran on to pick up the man on the ground, he was as light as a feather, I must have frightened or hurt him as he gave a loud scream attracting the bulls attention, he turned saw me and charged, my legs and brain went into overdrive, I knew I couldn't make it to the fence even on my own, I automatically made for the light then realised Bill would follow me in and squash me.

Feeling the ground shake as he gained on me I panicked its no good trying to swerve, I've seen Bill turn on a sixpence when the farm dogs tormented him, now I could feel his hot breath on my back, yards from the ramp I saw a gap under the machine, diving into the gap at speed and pushing the little chap in front of me was worthy of a circus acrobat the bull must have been only a yard behind us, he hit our refuge with such force I thought it would turn over, he bounced off and landed yards away in a tangle of horns and legs, he got up shook his head and charged us again. This time he almost lifted our shelter, in desperation I punched him on the nose as hard as I could but it made no difference, he only redoubled his efforts to get at us.

Bloody hell I'm trapped, if I run there's no trees around to climb, and the nearest hedge wouldn't stop a charging bull, I'd have to get out without attracting attention and run like smoke to get away. Smoke, that's it, rolling on my back I managed to get my fag lighter out of my pocket, I'd just bought one of the new type Ronson gas powered fag-lighters and had recently burnt a mates jersey, eyebrows, and half his hair demonstrating its flame throwing capabilities. Flicking the regulator to maximum I pointed it in the bulls direction and pressed the button.

WHOOSH, I don't know who was terrified the most, me or the bull, the sheet of flame in that confined space was more like an explosion, it burnt my face and almost blinded me. The bull snorted blowing out the flame and covering me in slime. He scrambled out from under the dome and charged off snorting and by the sound of it he was shaking his head from side to side.

I lay still for a while until my sight returned properly and Bill had disappeared over a rise in the ground, the little chap was still out cold and breathing heavily. I left him there and went to look at the other one. There was nothing I could do. Old Bill had made a mess of him, he had a large hole in his chest and his face was a squashed mess which was a pity I couldn't see what he looked like.

A sudden whirring noise stopped me from feeling sick at the gory sight, I turned and saw the square of light on our refuge getting smaller, the ramp was closing and all my kit was inside. My gun was my prize possession the thought of losing it made me furious, sprinting like mad I raced towards the dome and took a running dive over the top of the ramp. Luckily for me I landed across the padded seat next to the little chap who was supposed to be lying unconscious under the ship, evidently he had recovered and intended to take off. He squealed in terror at my sudden arrival, and I grabbed him before he could get away.

"Open the door," I shouted, but he didn't understand me even when I made the motion of the ramp going down, he just pointed to the screen in front of him, a thick black line was moving up to the top then he pointed up in the air, this must be some form of timer set for take off, the whirring noise in the background rose in pitch, as I let the little chap go the floor tilted throwing him across the control panel and he promptly fainted again. Picking him up I lay him on the reclining chair in front of the panel and checked him over, I thought he had a broken leg the way he kept fainting.

Looking for something to use as a splint gave me a good opportunity to look around the place, the inside of the craft was quite small, two couches were in front of a large screen and control panel on one side of the ramp, a room with a square window in it filled the centre, I tried to open the door to the small room, but there were no handles of any sort, leaving it for later I went back to see if my companion had recovered. He was still out cold and breathing heavily, to make him more comfortable I went to undo his tunic, this struck me as funny because the chap Old Bill had killed was wearing Earth type clothes.

Not finding a way to open his suit I left it in case he woke up, the only thing suitable for a splint was my gun, having no option I took it apart and tied the barrel to his leg, with strips torn from my shirt, in my search I hadn't found any sign of food or water which made me think our journey cant be long, perhaps this chap can sleep or hibernate till we get there, besides I haven't found a toilet unless its in the little room and I'm bursting.

Pressing the buttons on the control panel in desperation to see if the inner door opened only resulted in a small inspection panel opening, being so utterly desperate I must admit I couldn't help it and started to use the opening for a toilet before the little chap recovered. I doubt if I'd have made him understand what I wanted in time anyway.

= CHAPTER THREE =

A blinding flash and a terrible scream blocked out my conscious thoughts, I flew across the room, crashed into the wall and sank into a bottomless black pit.

Waking up and drinking the beaker of water offered, revived me into remembering back to this point, and brought all my memories flooding back. Now I know why this little chap is running around with my gun barrel tied to his leg, I can get on with my out of this world experience, knowing I can't do anything about it I settled down on the couch next to the alien and watched the pictures on the screen, at intervals a three dimensional picture of a planet would appear and a red dotted line swung around it, then fly off along a line of dashes. At first I thought they used a sling shot effect to gain speed, but the time interval between distances were regular which didn't tally with the vast different distances between stars. Watching our progress I realised they must have beacons on all the planets sending signals back to the spaceship, this means there had to be one on Earth, this could explain visions of celestial craft in the sky in the middle ages, these people have been visiting Earth for hundreds of years, was Jesus one of them who got left behind at some time, either accidentally or on purpose.

I tried using signs to talk to the alien but he wasn't interested in me, in fact he seemed pre occupied, he must be wondering how to explain my presence to his bosses, and why there was no bull on board, which made me wonder why send two frail little chaps across space to kidnap Old Bill and what would they want him for, the only reason I could think of was for breeding purposes.

Thinking about the bull brought back the memory of the chap he killed, something kept puzzling me, it was to do with his clothes, but for the life of me I couldn't remember what it was. Trying to reason it out I sifted through the clues what was he doing in Earth clothes, if he had come, from space he would be dressed like this chap next to me, he must have been a spy living amongst us gathering information to send back to his world, being occupied trying to sort things out and having missed a nights sleep I dozed off someone shaking my shoulder woke me from a lovely deep sleep In which I was dreaming I was hurtling through space in my own space ship exploring the galaxy, imagine my shock to find I was on a spaceship which was about to land, the alien who had woken me up pointed to the screen, the black line was now at the bottom of the picture, the same whirring noise I'd heard on take off sounded again, it appears the engines are only used for take off and landing the noise increased in pitch for a while and then died away, there had been no indication or even a bump to show we had landed.

Picking up my bag I eagerly made for the door then checked myself, what if the air wasn't breathable, what about gravity, what if there was a reception committee waiting with levelled ray guns, all these things flashed through mind and I realised it would be to my advantage to keep on friendly terms with the alien, so I turned and helped him to his feet as it was awkward for him to get off the couch with a splint on his leg, I supported him while he switched off the screen and dropped the ramp down, I nearly dropped him in my eagerness to look outside, I had passed the point of being scared there was a whole new world out there waiting to be explored and the prospects really excited me.

My first sight of this new world was a great disappointment it didn't look any different from Earth, the grass and a few nearby trees looked the same, although the sky was a pretty blue and there was a slight mist which made it impossible to see very far, I'd expected to land at an airfield or at least in a built up area but there was no one, In the near distance I could see a large building which at the time I thought was a hanger for the spaceship, but it turned out to be something entirely different, an object moving rapidly towards us glistening in the sun drew my attention and I waited expectantly to see what it was.

Believe nothing of what you hear and only half of what you see was my belief, but the thing approaching us was certainly out of this world, expecting to see a top of the range technological marvel I was confronted with the most laughable transport I've ever seen, a three wheeled platform with handlebars instead of a steering wheel, a seat for the driver two passenger seats and no engine that I could see, the trailer it was towing was no better, this also had only three wheels and a platform with some leather straps piled on it, no doubt this was to transport the bull, if this contraption was laughable the occupants were hilarious, the three of them looked like Chinese Hill-billies covered in dust they only wanted straw hats and clay pipes to complete the picture, Pulling alongside the lowered ramp they lifted it and hooked it onto their trailer.

I noticed the driver carried what looked like a spray gun with liquid in it, they completely ignored us and went straight inside the ship only to reappear moments later with a small box and my back pack, I hadn't seen the box before it must have been in a hidden compartment, while they were talking to the pilot I edged over to look in the box and was surprised to find it full of toiletries, Biro pens, some paper, a knife and a few more things I couldn't see without tipping them out, Without attracting their attention I picked it up and put it with my back pack with luck they'll think its mine and I could certainly do with the contents, I reckon they were souvenirs belonging to the dead chap back on Earth.

The newcomers finished talking, jumped down unhooked the ramp off the trailer closed it up, then motioned for us to get on board their actions surprised me they thought we had a bull with us which shows they can't have any radio contact with the ship, this didn't seem quite right a civilization that had space travel but no radio network, something funny is going on in this place, one minute they seem far in advance of us and the next minute far behind, bouncing along in the trailer wasn't very comfortable and I noticed my companion wincing every time we hit a bump, shouting for the driver to slow down had no effect, I'd been holding on to one of the straps which would have been used to hold the bull down, this gave me an idea, rolling the loose end into a ball I threw it into the wheel spokes it caught first time wrapped itself around the hub and jammed the wheel.

This brought the car to a juddering stop, the three chaps got off without saying a word, untangled the straps and went to get back on, catching hold of the man nearest to me I motioned for him to change places with the lame man on the trailer, but he pulled away to return to his seat, this really annoyed me and without thinking I picked him up and threw him on the trailer, I swung around fully expecting the others to come to his aid, instead of attacking me they just sat there waiting for orders, picking up the lame chap I placed him on the seat behind the driver, jumping on board I expected them to drive off, looking around I saw the driver looking at me as if waiting for orders, I was at a loss what to do so jokingly raised my arm waved a forward motion and shouted.

"Roll on the wagons," in true cowboy style we started of at speed, nearly falling off the trailer in amazement I thought what have I done, one minute I'm a prisoner and the next I'm in charge, now I know how the German prisoners of war working on the farm felt when their Polish guard went off with a local girl and they took his rifle back to camp and handed it in at the guardroom,

Being in charge didn't help me much when we arrived at our destination I was roughly stripped and shoved into a large shower cubicle the water was stone cold and full of chemicals it stung like disinfectant but had no smell, then the water changed it felt warmer and tasted normal, opening my eyes which had been clenched tightly closed against the stinging spray, I had the biggest shock of my life, the chap I'd rescued from the bull spent time cooped up in the spaceship and had tried to undress when fitting a splint to his leg was a girl, I had to look a few times to make sure There was no doubt about it, how could I have been so stupid not to have realised it before, she wasn't in the slightest way embarrassed taking her clothes off when an attendant came in to cut the rag tying my gun barrel to her leg, she never even tried to cover herself up, he looked at the gun barrel evidently not knowing what it was and threw it in the corner with the rest of my kit.

While still naked we were put on a large conveyor belt and passed through what I thought was an x-ray machine, I noticed they also put our clothes through which was a bit silly as we'd already contaminated the chaps on the trailer with Earth germs, after our treatment we were given towels to dry ourselves, of course I kept having a little peek at my companion just to make sure I wasn't mistaken yes she was definitely a girl, at first she appeared quite young, then I got the impression she was in her early twenties, the attendants brought us new clothes to put on.

I was pleased to find mine fitted reasonably well even my shoes were the right size, much different to when I joined the Air Force, they issued me with odd sized boots, and I had to steal one from stores to make up my pair, I expect every new intake of recruits would have to do the same so in the depths of time the chain will end with one pair of odd sized boots left on the shelf, the fasteners on my new boots were the same as the ones on the girls clothes. I reckon if they were fitted to the soles of boots we could walk upside down on ceilings or better still used on runner training equipment would build up athletes leg muscles by making it harder to lift their feet

I was still playing with my boot fasteners when two attendants came in to escort us to a reception committee which consisted of four men and a woman, she sat in the middle and it was obvious who was the boss, this was the first time I had met someone in authority, she was dressed the same as everyone here except she had a solid looking gold band around her neck, Which I assumed was a badge of rank, the men were dressed the same but I noticed they all looked older than the ones I had met so far, they kept looking at the woman waiting for her to start the proceedings.

I was interested in what was about to happen, she motioned for me to bend, I thought this was for me to show respect, instead an attendant stepped forward and snapped a thin band around my neck, I heard the crack of an electric spark and a wisp of smoke curled past my face up towards the ceiling, kicking myself for being caught off guard, I felt it expecting it to be a restraint of some sort but there was no chain attached to it.

My service training had included interrogation tactics used by the Russians and the Chinese, so I had a good idea what to expect, "What right have you to come here uninvited," Christ I nearly wet myself, I'd never expected to be spoken to in English, I looked around to locate where the voice was coming from, "Well answer me," The voice spoke again and it came from the band around my neck, that's what its for it's a translating device, I could make a fortune with this back on Earth, pre occupied with these thoughts I was slow in answering.

"Come on stupid, we haven't got all day," Her attitude stung me into answering, then I stopped myself, she was using the Chinese method which is based on people only speak the truth when really angry, OK I thought lets beat her at her own game so I shouted back at her, "Shut up you mangy old cow I have as much right to be here as you, I saved the life of your pilot, brought back a valuable spaceship and had to board it to save my own life, you should be grateful to me and show your appreciation by showing me a little respect otherwise your superiors will have something to say."

She was shocked at my outburst and at a loss for words, then she changed her tactics and tried to be friendly by asking what I did for my obligations, I suppose that's what they mean for work, so I told her I worked on the latest jet fighter aircraft, and research into anti gravity devices, I'd noticed none of the men had translators on which seemed a bit silly, why have them here if they can't understand what I'm saying, she held up her hand to silence me, and touched the band around her neck which switched off her translator, then spoke to the other committee members in her own language.

Then she turned angry and started shouting but she hadn't turned her translator on so the words were lost on me, I laughed and this really upset her turning it back on she repeated her tirade, "Your telling lies we know your worlds ignorant of gravity" she screamed, I certainly got the message.

Realizing she was going to get worse I decided the best defence is attack and brazen it out, so I shouted back, "If your spies had been any good, they'd have found our ancestors using anti gravity devices thousands of years ago, natives of Easter Island moved great stone statues by using resonance tubes which produced atomic frequency oscillations, these cause electrons to wobble in their orbits, altering the direction of the pull of gravity, and by changing the pitch they actually made the statue tilt each way and walk, if they got the pitch wrong, the statue toppled over and couldn't be lifted again, they are still lying there for anyone to see, besides any child knows it's the reactive braking effect on outward moving electrons that make atoms attractive to a larger mass, that's why denser atoms weigh more."

Now I'd got into my stride, I thought a few truths would give them something to think about, I told her we have well documented evidence in a book called the Bible available to anyone, which describes in detail how our ancestors destroyed a town called Jericho by using a similar type of device called a trumpet, and if she wanted more proof she could send her spies to Baalbek in the Middle East where they would see for themselves. Large slabs of rock big enough for the Romans to build a temple on, they were quarried and moved many miles to their present position, one large slab had been left broken in the quarry, which is proof of their original resting place, it crossed my mind to bring the Pyramids into my story, but changed my mind, these people may have helped to build them, using their own methods for moving blocks of stone.

Waiting for a break in the talks I asked for something to eat, she beckoned to an attendant, who went away and returned with a tray of biscuits and a jug of water, I took two biscuits turned and handed the tray to my companion who was standing behind me, so far she had taken no part in the meeting, seeing the concern for my former pilot the woman switched off her translator and spoke to her in their own language, then she spoke to the attendant again.

He left the room and returned with a gold translator ring which he put around my companions neck, this time I saw the spark and the rising column of smoke as the two ends welded together, this puzzled me as sparks are fatal to transistors, and usually burn them out, I wondered why they were welded on, I could understand if they were for tracking prisoners but for a high government official to have one doesn't make sense, I was about to ask her, when she got up, ignored everyone and walked out of the room, by her action I assumed the meeting was over. The committee filed out after her, leaving me wondering what had been discussed about my future, was I a prisoner, a slave or a hero for saving a spaceship I was at a loss what to do, until my companion caught my arm and said, "Come on I have to look after you," this made me laugh coming from a wisp of a girl of twenty something with a broken leg.

= *CHAPTER FOUR* =

Then I noticed she hadn't put the splint back on, perhaps the X-ray machine had shown it wasn't broken, I also realised she had spoken in English, this was great now I could ask her the thousand and one questions I wanted answered, like what's going to happen to me, she said that as a punishment for bringing me to the planet she has to look after me for a while, and show me where to get my supplies, she is also responsible to make sure I don't cause trouble,

By her attitude I don't think she was very happy with this arrangement but told me later it was a much lighter punishment than she would have got from the Founders, at the time I didn't take any notice of her reference to them, I was more concerned with plans affecting me, but she said nothing had been decided about me it would be discussed at a higher level

In the meantime I was to be isolated from the dwellings, and the obligations awarded me for saving the spaceship would ensure me enough food for the rest of my life, this shook me I hadn't thought of spending the rest of my life here, the novelty of being the first Earth-man to set foot on another world would soon wear off, then I'd become bored and end up like Robinson Crusoe on his desert island without his man Friday, unless I can use my manly charms and persuade my companion to stay, she wasn't a beauty but even a Kangaroo would begin to look beautiful after a few years, living on my own.

First she said we collect supplies, then I have to take you to the hut and look after you, this certainly suited me so I didn't raise any objections, we loaded up with plastic bottles of water and packs of the thick biscuits we had eaten at the meeting, When I asked why just biscuits she said eating food made you old and gave you diseases the biscuits contained all the vitamins and minerals they ever needed, looking around the place for anything useful I found my kit and rifle barrel in the shower room, the stock for it was still in the spaceship, where I'd left it, if the gun had been complete I'm sure they would have taken it off me, given a suitable piece of wood and enough time I can soon make another stock for it,

With a full load we set off on the car we came in, she managed to drive easily enough it only had the handlebars, and one pedal, I thought I'd have plenty of time to study the car later, so I concentrated on looking at the countryside, this consisted mostly of scrub-land with stunted trees dotted about, we passed a couple of dew-ponds with rabbits drinking from them, this was interesting as I hadn't seen any clouds, just a light mist and wondered if it ever rained, I was so engrossed in looking around I didn't notice the building until we stopped outside it.

Talk about a surprise this was a big one, of all the things to find on a new world I didn't expect to find a log cabin, well it looked like logs until I had a closer look, it was made with plastic pipes about five inches in diameter and twenty feet long, it was an amazing building the pipes fitted together so closely it was wind and water proof, the roof was made with pipes cut down their length and every other half turned over, walking around the building I marvelled at the construction, pushing open the door, I had to wait for the dust to settle, before going in to open the shutters, inside it looked like someone had left in a hurry, both rooms and the storeroom at the back were covered in broken clay pots, the shelves had rotted away and the pots had smashed on the stone floor, some pots had survived but the contents had dried up long ago, going back out to the car I asked the girl who built the place,

"The other Earth-man built it a long time ago," she said, I thought I was immune from shocks but this one stopped me in my tracks, “What other Earth-man,“ I asked,“ “ He lived here a long time ago, But I know all about him, he was part of my history studies, This is the story she told me.

= *CHAPTER FIVE* =

About two hundred years ago, our Earth mission landed in North America, on previous visits the natives had been friendly, but this time they attacked the crew killing one of them, the other man got away but he was badly wounded, he was found and cared for by a white man who had seen the ship arrive, to escape the Indians he carried the injured man back to the ship, only to find two men guarding it, the white man shot one and fought with the other throwing him outside, this gave them time to start transition.

Arriving here our ancestors thought the white man had attacked and wounded our man, but he recovered in time to stop him being thrown in the crushers, instead he was rewarded for saving a valuable spaceship he didn't want any obligations, instead he asked for materials and help to build this place.

He lived here alone for a while then got lonely, and made himself a nuisance by going to the mines and upsetting the order of things, he gave the miners a liquid which made them unfit for work, in view of his rejection of our obligations it was decided to return him to Earth which repaid our obligation, and it was agreed that if he was returned without his weapons to the place he came from the natives would kill him, saving us the problems of Earth people finding out about our visits, the mission taking him back, reported erratic behaviour by the earth-man, with no translator he tried to tell them by signs the trees were taller and the roof of his log cabin had fallen in, our historians deduced a time slip had occurred which explained why previously returned Earthmen had never reported back.

Her story astounded me not only about the other Earthmen and the time factor involved, it was the fact that if he'd been returned why cant I be taken back, another thing she said, it was decided to take him back, sounded like a committee or a governing body of some sort make the decisions and the workers carry them out.

The more I think of this place the more it resembles a bee colony, they only eat biscuits and drink water, they get no pay for working, and the sexes have no interest in each other, I wonder if my companion or jailer will soften towards me now we are living together, so far she hasn't shown any interest in what I've been doing, I wanted her to help by using her own initiative, its really annoying having to instruct her every time, lighting a fire was a good example, I told her to put the rubbish on to get rid of it, I don't think she had ever seen a fire before, she was so mesmerised by the flames, she kept piling it on, being bone dry it flared up and if I hadn't thrown the pot of water on it, she would have burnt the place down., shouting at her for being so stupid didn't have any effect, at times like this I wondered how she ever managed to become a spaceship pilot, although none of the people here seem very bright, I wonder if they are all on drugs.

Clearing out the place made me feel like a burglar intruding into someone's personal possessions, the previous owner had left everything behind, some of the utensils would come in handy, his metal pans interested me they had been beaten into shape, and the furniture was well made, overall it would make a nice cosy place to stay, all I wanted now was a nice hot meal, I hadn't had one for ages and I know how to get one.

While outside collecting firewood I'd seen plenty of rabbits running around, selecting a stout stick I set off out to get one, chasing them was futile, they were like hares and easily evaded me, in the end as the light was fading I just stood still and waited, soon one came hopping by, Whack I gave him a good hefty clout which must have broken his back, he squealed in agony, as I bent to pick him up, Crash, something heavy thudded into my back pitching me flat on my face, it grabbed the squealing rabbit out of my hand and bounded off into the darkness.

Lying there winded probably saved my life, ghostly shapes hurtled over me in pursuit of the noise, their snapping and snarling terrified me, Bloody hell wolves, they must have been stalking me ready to pounce, then got distracted by the squealing.

While the sound of fighting faded into the distance, I got up and made my way carefully back to the cabin, Boy was I relieved to hear the door close behind me, I'd been scared stiff which showed, by me catching hold of the alien, and shouting at her "Why didn't you tell me about the wolves," she just shrugged her shoulders and said," You never asked me," her indifference really appalled me, and I realised I don't know much about this place, so far I'd only seen rabbits and wolves, still feeling a bit scared I asked if there were any more dangerous animals I should know about; she hesitated at first then told me about the Outlaws, she said.

"They never speak about them, as they do naughty things to anyone they capture," I burst out laughing at this funny remark, it certainly aroused my curiosity, I wanted to know who they were and where they came from, but no matter how much I asked she wouldn't say what the naughty words meant, even when I reminded her she was supposed to be looking after me, if I got hurt by something I didn't know about she would be held responsible, this still didn't loosen her tongue about the naughty bit, but she told me some more about the Outlaws.

It appears in olden times they were sent here to keep the number of dogs down and supply them to the food farm, this shook me, I couldn't believe what I was hearing, you mean to say the biscuits we have been eating are made with dog meat, if I'd known I wouldn't have eaten them, " then you will have to go hungry "she snapped back, "that's all we eat, they are made in the food farm and contain all the vitamins and medicines we need for a healthy life, much different to the problems you have."

"We have no plagues and illnesses like you have on your planet, this was the longest time she had spoken and I wanted to find out more, until a nagging thought came into my head, "what do you do with the bodies when someone dies, do they go into the biscuits," "no" she said "they go into the crushers for fertilizer," that was a relief to hear but she carried on, "we cant use human remains in the biscuits they live too long and absorb too much radiation, their readings are too high it would pollute the mixture," silly me thinking they didn't use them on compassionate grounds.

Getting undressed for bed was a little embarrassing for me, but it didn't bother her, she came over and touched the band around my neck and then spoke, I couldn't understand her, she had switched my translator off, taking my hand she showed me how to switch it on and off, I suppose it was to save the battery power, stretching out on my bunk bed which I'd covered with small branches to make a soft mattress, I wondered about my companion, I didn't know her name, where she came from or even if she had any children, in fact I didn't know anything about her.

All the time I'd been in her company I hadn't asked any personal questions, perhaps she was thinking the same about me, calling out goodnight didn't get a response I'd forgotten my translator was turned off, she wouldn't know what it meant anyway, falling asleep to the sound of the fire crackling sparks made me dream I was back on Earth .being chased by wolves and women.

Waking up to the sound of a dog barking in the distance, lying in a nice warm cosy bed half asleep, I felt relieved to think it was all a bad dream, when I became aware of a sharp pain in my neck something was causing me trouble, reaching up to remove it my fingers encountered a metal band, Bloody hell, I leapt out of bed as if I'd been shot, it's not a dream its true, now all the events of the last few days flashed through my head, just like a drowning man's life flashes through his mind, a voice spoke but I didn't know what it said until I turned on my translator, my friend had also jumped out of bed and wanted to know what was wrong,, looking at her standing there without a stitch of clothing on, and not showing the slightest embarrassment startled me, I swear to this day that all I could think to say was "Good Morning miss, I'm sorry but I didn't get your name, last night," she gave me a funny look, and said her name was Worra which I thought was very appropriate, looking at her she was certainly a sight for sore eyes.

While building up the fire it crossed my mind to set snares to catch rabbits, it wouldn't make any noise to attract the dogs, I don't fancy tangling with them again, or eating biscuits if they're made from the bodies of dead ones.

Having time to think, I reasoned if they make the biscuits here, they must grow all the raw ingredients, Worra had mentioned a food farm, now I asked if she knew the way to it, "Yes" she told me "She worked there for a while on one of her obligations, she had to travel to food farms on field studies for the breeding programmes," this information I mentally filed away for future reference, at present I was more interested in finding food, Collecting my kit a few small pots and a large metal pot, we piled it all on the car and set off for the food farm.

Worra was driving, I was riding shotgun, there's no way of knowing what we would encounter, the scrub-land was in poor condition but it gradually got better. Now the morning mist had cleared, In the distance I saw rows of taller trees imagine my surprise to find they were apple trees, and better still they had a nice crop of fruit on them, stopping to investigate I found they tasted lovely so I picked a few and filled a couple of small pots,

As we got further into the miles of orchards, I noticed the trees changing, gradually we went through the growing seasons, there were even trees with blossom on, It looked like they had all the seasons growing in one orchard, this meant they had fruit all the year round, I expect this form of cultivation would be used on all other forms of food production, this place is a veritable Garden of Eden, but after thinking it was so wonderful, I realised, we have the same thing on Earth, we get fruit all year round, its always summer somewhere, looking around I noticed there were no fences or protection from bad weather or sun shine, it must be this sort of weather all the time, without any drastic temperature changes, water comes from the heavy mist that hangs in the air in the mornings, there are a few dew ponds which the animals drink from, I hadn't seen any sign of rivers, the only other interest was a few workers toiling away, I asked Worra why they were totally ignoring us, she said.

“They have their quota to do, if they don’t complete their obligations they have to work in the mines“, more surprises, no one had mentioned any mines asking Worra she told me the mines were the only reason they were here, her world had used all its resources over thousands of years and now had to find other supplies, this planet was rich in Uranium,, which they traded with the home world for goods and technicians to run their power stations, She also dropped the bombshell that she would be going home on the next transport due tomorrow, this alarmed me I’d be lost without her, I mentioned this but she said she had no option she had to go, this was a blow I hadn’t expected, just when pieces of the puzzle were falling into place, this planet is only a mining satellite, all the machinery and people came from the home world, It also explains why they have space ships and higher technology for such a small world, no wonder they appear to be so backward.

My daydreaming was cut short, by our arrival at the food farm, the buildings were a novelty they were all the same shape and size I asked Worra why they were like this, but she didn’t know. Examining them I marvelled at the simplicity of their construction, they were just fourteen foot diameter discs made of quarter inch thick plastic, two were solid with small slots near the edge they made the floor and roof, the other two were dissected into panels radiating from the centre, one disc lay on the floor disk with the dissected panels lifted and pointing up like a kings crown, these would be held upright by the tips going through slots in the outside circumference of the roof and floor discs, the same had been done with the other two discs, these were inverted, and placed on top, so the spikes interlocked and formed the walls, panels could be left out for doors or windows, looking around them reminded me of the prefabs on Earth, small cheap mass produced bungalows built during the war, I could imagine people living in these, they would be ideal in Africa for living in or used for storehouses.

No one interfered while we helped ourselves to a variety of items, some of the things I’d never seen before, so I left them alone, I wasn’t sure of the meat, whether it was dog or rabbit, so I decided to catch my own in future, not being sure about Worra staying with me, I only collected stuff for myself, she only ate biscuits anyway so it solved that problem, having collected enough fruit and vegetables to last us a while we set off back to the cabin.

On the way I racked my brains to think of a way to keep her here; still thinking about the problem as we neared the cabin I realised it had answered itself, Worra had watched me eating apples with relish, and being a typical woman had eaten one or two herself and enjoyed them, it was a new experience for her as she normally lived on the biscuits which contained all the goodness and vitamins they required, her stomach not being used to raw apple got upset; by the time we stopped she was writhing about in pain, for a while I was at a loss what to do, it was no good trying to kiss it better.

= *CHAPTER SIX* =

Then I remembered my mother always gave us a cup of warm water to cure our aches and pains, carrying Worra into the cabin I lay her in on the bed gave her a drink of warm water and told her it was the proper way to take away the pain, and she would have to stay in bed for a couple of days, or the pain would return, she spent a restless night, but appeared a lot better next morning.

At midday we heard the whine of a car approaching, the party had come to collect her for her homeward journey, they had no translators, so I didn't know what they were saying, two of the men went to pick her up to carry her out, but when she was violently sick they soon dropped her, they couldn't get out of the room fast enough, it couldn't have worked out better if I had planned it, although I thought it was a little suspicious she had appeared a lot better this morning.

My suspicions were confirmed later, when I saw two apple cores lying in the fireplace, looking after Worra I didn't see the men go, and it was only later I found they had taken the car with them.

This was going to cause me problems I wouldn't be able to travel far, there was no way I was going to camp out with packs of hungry dogs running around, it also meant I couldn't take Worra with me, she still limped a bit, I asked if their was a doctor or medical centre at the mine, of course I had to explain what a Medical centre was," No she said there's no need for one, minor injuries heal themselves and serious injured ones are terminated and put in the crushers for fertiliser, she said all this without a flicker of emotion, just a matter of fact statement, no wonder this community is self supporting nothing wasted or thrown away just like an ant colony, I'm surprised they don't eat the bodies, although I suppose they do in a way, no different to us burying dead bodies back on Earth; only we have worms to recycle ours.

I asked if she would have gone in the crusher with her bad leg.

"No" she said "it never affected me piloting a spaceship, we haven't many pilots or spaceships left and we cant build any more, that's why you will be well looked after, in return for saving one," "Our ancestors squandered our resources on journeys to distant planets, by the time they found out most of them were barren rocks it was too late, the colonies set up on other worlds were left to look after themselves.

Earth is one the Founders allow us to visit, and report on, it is the only one here to survive on its own, the other worlds nearby were seeded with the best brains but they degenerated into tribal wars, or died out with disease, or famines, It may have been that your world was more fertile, also the tribes were set up farther apart, by the time they made contact they were civilized and traded with each other, now you are entering a time of peace all your major wars are over, you have learned from the age of conflict no one gains by fighting."

Of course I burst out laughing, and said if she expected me to believe such rubbish, she should make up better stories than these .for a start the Earth I left is even now at each others throats, Russia and America are arming themselves with enough atomic weapons to annihilate each other as fast as they can, the Israelis and Arabs are on the warpath, and the Chinese and United Nations, are fighting in Korea as we talk", "that's true "she said. "You must have your little squabbles it gets rid of aggression, but you wont be allowed to have a nuclear war, that will upset the balance," this evoked a bigger laugh from me, so far I'd had a slight interest in her story, now I knew she was trying it on, "Your trying to tell me your founders or whatever they call themselves dictate what happens on Earth," her answer startled me.

"Of course they can, don't forget your ancestors came from our world long ago, and their brain patterns are still linked, don't you ever wonder where your stories of gods and religion came from."

"Our founders used it to get the pyramid beacons built and cause wars through the ages, the Founders even got humans to pretend to be sons of god and start religions of their own, this caused conflict which diverted the tribes from building towers to join our ancestors in space," she realised I was laughing at her and redoubled her efforts to convince me.

"You must have heard of Joan of Arcs voices, and angel messages. in terrible times, you always get someone to put things right, your latest man Churchill was guided by us" a lot of what she was saying made sense, but by getting her angry she will tell me more, "Come off it" I said "you don't think I'm going to believe such a load of rubbish, how would a lowly space pilot know so much about high council business," "that's easy she said.

"My partner lived in the commune where the Founders visit and he was obliged to serve them, they appeared more often in old times, we also had our little disagreements, but now the problems of setting up this colony have been sorted out they don't come so often, when I was studying your history and said I couldn't understand how Earthmen could leave their homes and family to go off and kill people like themselves, my partner told me all about the founders plans for Earth."

Trying to get some sanity back into our conversation, I ridiculed her story, and said, "You must think I'm an idiot to believe you, even if your partner came and told me, I wouldn't believe him", "There's no fear of that," she snapped back.

That bull of yours saw to that back on Earth," It didn't sink in for a minute then it hit me,

"You mean to say the chap the bull killed was your partner, why didn't you tell me before," "there was no need" she said

"He was dead and there was nothing we could do about it, I suppose we could have brought him back but he would only have gone in the crushers," getting used to the callous indifference of the people here still appalled me, but every thing is so practical, it's their way of life, sentiment doesn't come into it.

Now was my chance to find out what happened on Earth,"What did you do to our bull to make him so mad, he was normally as docile as a lamb," she took her time answering, as if she was trying to think of an excuse, then decided to tell the truth, "We were trained to control cattle and had no problems before, we went through the normal routine, but when Axel, that was my partners name, sprayed the dope mixture up the bulls nose to make him sleepy he sneezed, Axel caught hold of the nose ring to hold him steady to try again, when the bull swung his head sideways the horn ripped into his chest, the scent of warm blood sent the bull into a frenzy he tossed Axel high in the air then gored him when he landed,, I tried to distract the bull even though I knew my partner was dead, but I was knocked over and hurt my leg."

"I was trying to get back to the ship when you picked me up and I fainted., I remember seeing a big flash of light, then you disappeared, so I crawled into the ship and pressed the emergency transfer, I was terrified when you landed by me, I thought I had pressed the auto destruct by mistake," she stopped suddenly, , "Whatever made you jump aboard, you must have known you would never return, Van Winkle was taken back because of our obligation laws and we had to get rid of him."

Worra was really letting herself go, this was the first time she had called the Earth-man by name, while she was in a friendly mood I decided to take her with me, I had to set a snare to get a rabbit for dinner, and a piece of wood for my rifle butt, being occupied chopping a tree down with the dogs and Outlaws around, didn't appeal to me, she could be my lookout, the small axe I'd found in the storeroom would have to do, although I thought it would be better for fighting, We didn't go far before I found a decent sized tree cut it down and trimmed a piece to the size I wanted, for a bit of fun I cut a piece of wood split it and made a crutch for Worra, showing her how to use it was a laugh,, she was like a kid with a new toy, I think it was the first time she had ever been given anything.

Back at the the cabin I found my snare had caught a rabbit which I soon had his jacket off and he was boiling away merrily in the cooking pot, while waiting for it to cool I started clearing out the rubbish In the storeroom, the light in there was very dim, and I was worried in case there were any rats, taking a burning ember from the fire I used this to examine the place and soon found a bandoleer hanging on a peg, five of the pouches had black powder in them, taking a pinch I threw it in the fire Whoosh, a bright flash lit up the place.

This was a bit of luck, gunpowder will came in handy all I want now is a musket to use it in, crossing the room to open the shutter, I turned and as if in answer to a prayer an object appeared hanging on two pegs above the doorway a musket, I couldn't believe it, not until it was actually in my hands, now I know the owner was forcibly ejected, he would never have left his most treasured possession behind, looking for musket balls was a disappointment, I couldn't find any, not until I'd given up and was picking rubbish off the floor. Then I found dozens, they must have been hanging up but over the years the leather pouches had rotted and the heavy balls had fallen through .I couldn't wait to try it out, ripping off a piece of my shirt I went outside poured a bit of gunpowder down the barrel rammed a bit of rag in and the ball last, using a match head from Axel's box in the priming pan, I aimed at a nearby tree, shut my eyes and pulled the trigger, the bang and scream sounded at the same time, Worra had been outside collecting firewood, For a panic stricken moment I thought I'd shot her, I don't know where the bullet went, but I know It had terrified her, of course she had never heard a gun shot before, running after her into the cabin I found she was all-right, but my dinner was boiling over, my rabbit stew was delicious even though there was no salt in it, Worra tried a bit but I think she was remembering the pain of the apples, talking to her after the meal I asked if there were any more buildings.

"There are" she said.

"But we don't go near them, many engineers were killed trying to enter them," this sounded funny, when I pointed this out, she said "The buildings were here when the planet was colonised," this certainly got me interested, I'd love to find out what they contained, shut up for hundreds of years built by an unknown race, they were certainly worth exploring, asking for directions she tried to stop me going by saying "it was too dangerous," seeing i was determined she told me "they were near the hills we had seen on the way to the food farm," deciding to set off in the morning I collected my kit and the musket ready for an early start.

Excited at the prospects of what I would find I set of as soon as it was light, the usual mist was hanging in the air cutting down visibility, it was only after it had cleared that I realised how dangerous it could have been, I wouldn't have seen any attackers stalking me, taking notice I wouldn't be so silly in future, now I could see the hills it wasn't long before the building came into view, getting closer it looked like a massive flat roof structure with a lump in the middle, thinking of the hump I'm sure it was a shot tower for making ball bearings, if that's the case it had to be a factory but for making what, busy thinking, I soon stood before the massive doors the whole place was a terrific size, I walked to the sides to look for windows or doors but couldn't see any.

Checking over the doors I marvelled at the workmanship, they were made in two halves and you couldn't get a match stick in between any of the cracks, there were burn marks on the centre of the doors and a few dents, a one inch hole was in the middle of each door, where the handle would normally be, on one side door post was a square panel with four black squares on it, I decided the holes must be the best place to concentrate, first I had a good look around for a handle to wind the doors open, the people who left this place must have left a key somewhere close, the obvious place would be on the roof or under a carpet, but I couldn't see anything for use as a ladder or any sign of one being buried, I even looked for writing or signs in case they had left instructions where it was, to say I was disappointed was an understatement.

What I'd seen of the place made me more determined than ever to get in, now I had an idea what I was up against, I would come better prepared.

Noticing it was getting late I decided to leave, or I'd be travelling in the dark, and that's the last thing I want to do, I'd been stalked by a pack of hunting dogs before and don't want to go through that experience again, hiding some of my equipment in nearby bushes to save carrying it back I set off for the cabin, on the way I had the feeling I was being watched but no matter what tricks I tried I never saw anyone, or anything, Worra had told me there were only dogs and Outlaws who were dangerous, I had already met the dogs so there was only the Outlaws to worry about, to let them know i was aware of their presence, I loaded the musket with half charges and a handful of dirt then fired into the trees, this made an impressive bang, and a cloud of smoke coupled with a shower of leaves.

Arriving back at the cabin without any hindrance I was relieved to see smoke coming from the chimney, when I called out Worra opened the door, she had heard me shooting and thought I was being attacked, but as the bangs got closer she knew I was al-right. Back in the safety of the cabin made me realise the danger I had left her in, the Outlaws could be watching the cabin all the time, and I had left her with no means of protecting herself, dragging the previous owners tool box out of the store room, near to the fire, I selected a large knife, and started to shape the piece of wood I'd got earlier to make a new stock for my rifle, sat there whittling away, I noticed some letters burnt into the lid of the toolbox, curious I turned the box to read them better, RIP VAN WINKLE, that sounded nothing like Worra said it was, I had to read it again, I'm sure I've heard that name before, Of course, it's in a story about a chap in North America who disappeared in the seventeen hundreds and returned fifty years later, still looking young, I think he was burnt as a witch, because of the stories he told.

Trying to remember what I had read about him, I stopped work to concentrate, instead I became aware of a noise in the distance gradually getting closer, giving it my undivided attention I soon realised it was a pack of hunting dogs in full cry, after some hapless creature, expecting the chase to go hurtling past and recede into the distance, I was stung into action when CRASH a heavy body thudded against the door and I'm sure I heard a human cry out amidst the sound of snapping and snarling, gripping my piece of wood I flung the door open, a vicious fight was taking place outside, a shadowy figure was pinned against the wall with a pack of dogs snapping and snarling and fighting each other in their mad blood lust to overcome the lone fighter, by the amount of fur flying about I had no time to lose, shouting out I jumped into the fray lashing out at the snarling pack, bowling dogs over left and right,

My sudden attack from the rear had the desired effect, the pack drew back, giving me enough time to grab the body by the fur, and drag it towards the doorway the dogs attacked again, one landed on my back sending me sprawling into the room, panic stricken I reached over my head caught him by the loose fur on his throat, turned and threw him bodily out through the open door, he landed among the dogs fighting to get in, this triggered a ferocious fight to the death, as each dog attacked its neighbour, fascinated I watched through the open door as they tore lumps of flesh and fur off each other, I'd seen dogs fighting before but nothing like these, it looked like the fighting would never stop the noise rose to a crescendo then receded into the distance as the running fight moved away.

Kicking the door shut I found Worra lying unconscious on the floor behind it, she must have been about to bar the door when I barged into it with the dog on my back, by the lump on her forehead she was going to have a right bad headache in the morning, picking her up and laying her on the bed I turned to look at the other body, in the poor light of the lantern it looked in a bad way, lumps of fur were hanging off and it was covered in blood in places, stood there looking down at it I was at a loss what to do, if I approached it might attack me, but I just couldn't leave it there.

Then I had an idea it had been running to get away from the dogs so it must be thirsty, crossing over to the table I filled a mug full of water put two biscuits on a plate and took them over to the inert form, placing them on the floor I sat back to watch, Boy was I relieved when a bare arm emerged from the fur and took the water to drink, now I could see the fur was rabbit skins sewn together to make a cloak with hood attached, I also noticed blood trickling down the arm this reminded me I had a few bites of my own to attend to, pouring hot water into a bowl I rolled up my trouser leg and started to bathe my wounds with a pad of cloth.

Suddenly I saw movement by the door, alarmed I jumped up but I was too late, the figure took the cloth off me and threw it on the floor, then it produced a small pot from inside its cloak, poured a small amount of liquid on its bites, rubbed it in with its other hand, and did the same to me, then it extracted something that looked like a piece of cake from inside its cloak broke a piece off and ate it, the smell was vile and when it offered me a piece I pushed the hand away, it pointed to my bites.

Then shoved it into my mouth holding its hand in place to make sure I swallowed it I looked up hoping to see a face but it was in shadow, so I made the motions for it to pull the hood off, instead the figure took my hand and before I could stop it, pulled it inside her cloak, and pressed it onto a warm breast, talk about being embarrassed it was such a shock. Her next move now I know it was a she really flipped me, she leaned over and thrust her hand down the front of my trousers, To gently squeeze my manhood, well I was speechless, it's a good thing we don't have greetings like it on Earth, otherwise we would all end up either in prison, or being greeted by everyone we met.

A groaning noise from the bed brought me back to reality, Worra was sitting up rubbing her head, her eyes widened in fear, as she looked past me at the grotesque figure standing behind me in a misshapen fur coat, she squealed in terror and dived under her bedclothes, no matter how hard I tried she wouldn't come out, in the end having no option I pulled them off her, but she clung to them so tightly we and the clothes ended up in a tangled heap on the floor, the stranger burst out laughing at the sight of us wrestling on the floor, this did the trick, Worra suddenly stopped fighting to compose herself and glared at her.

"Come on I said don't be silly ask her name and why she was being chased in the dark," Worra grudgingly gave in and translated for me she said the strangers name was Valda and she had been sent to spy on me, but when I had pointed my stick which made a loud noise in her direction she had run away and hid until it was nearly dark, she thought she could reach her village before the dogs started hunting, but the dogs had picked up her scent when she still had a long way to go, in desperation she had headed for the cabin, hoping to climb on the roof to escape the dogs, I asked Worra to ask if she would take us to her village in the morning.

“No,” she said, “I owe you many obligations but I will have to ask our head man first, we are afraid of strangers, our folklore tells of the stranger who lived here long ago, who had a stick like yours which made a loud noise and killed people, but you saved me from the dogs so you are not a bad man,” now I’m in the Outlaws good books I asked Worra to find out what was the nasty things they did to anyone they caught, she hesitated then did as I asked, Valda looked at me shook her head and burst out laughing, Worra said it was a secret, the way they both started chatting and looking at me made me feel uncomfortable, Worra had turned her translator off so I couldn’t make out what they were saying, anyway she can tell me all about it tomorrow, showing Valda she could have my bed she examined it then wrapped her cloak around herself, lay on the floor and went to sleep, just as if the traumatic events of the last few hours happened every day.

I lay awake for a while pondering over the things that had taken place, and the unreal things that were going through my mind, I mean there are no children, or insects, although there were plenty of bees up at the food farm, there are no birds either, I hadn’t noticed that before, it seemed that everything here was put here, the dogs were the odd ones out, perhaps they were put here to control the rabbits, and the outlaws were put here to control the dogs which seems a much better system than ours, we have to pay to have our criminals locked up, then guard them for their length of stay, whereas here everything appears more common sense even the weathers so sensible it must be man made.

= CHAPTER SEVEN =

The morning mist keeps it moist and so far there doesn't appear to be any variation in the seasons, sleep must have overcome me at this point because the next thing I remember was waking up to the delicious aroma of meat cooking, looking outside I saw our new lodger tending a fire with my stew pot hanging over it, on closer inspection it looked like rabbit stew, this was more like it I've got the life of Riley with two women waiting on me.

I went back inside to finish dressing when a shout warned me something was wrong, dashing out into the sunlight another shout alerted me just in time, a man was running towards me, I stepped aside and tripped him as he went past, he went down in a tangle of arms and legs, I jumped on his back put my around his neck and pulled his head back, this had happened so fast I did it without thinking, looking around I saw three more men watching me, they hadn't come to their friends aid, then I saw why.

Valda was threatening them with my large cooking knife, she came over and slapped the helpless man I was holding hard across the face, she followed this with a barrage of vicious slaps that rocked the chaps head from side to side, I couldn't just sit there and be a party to such an attack on a helpless man, loosing my hold on him, I helped him to regain his feet while he was trying his best to fend her off.

I noticed a knife on the floor, before I realised what I was doing I picked it up and handed it to him, it must have been all those Hollywood Indian films I'd seen that made me do such a stupid thing, the chap stood there with the knife in his hand looking puzzled he couldn't understand what was going on, the girl broke the deadlock by slapping his face again, I caught her arm to stop her, this resulted in a burst of laughter from the other men, looking back at the event I couldn't help laughing myself, there I was fully prepared to fight for my life against odds of four to one, instead I was stopping a young girl from slapping my assailant, and true to life she turned on me, she came at me tooth and nail her ferocity sent me staggering, for someone with such a small build she bowled me over, the attack had been so sudden I struggled to defend myself, in desperation I wrapped my arms around her, she tried to bite me so I gave her a light head butt, this knocked her out and she slumped to the floor, whirling around fully expecting the rest to come to her aid, I didn't have to worry the four men were lying on the ground holding their sides convulsed with laughter, now I was the one at a loss to know what to do.

The girl was lying still on the ground, so I got a mug of water and splashed some on her face, then when she showed signs of life I held it to her lips after taking a few sips she opened her eyes and looked at me, I wasn't sure what she would do, so I waited awhile before helping her to her feet, she gave me a funny sort of half smile, before returning to the cooking pot.

The men had lost interest in the fight and were helping themselves to my breakfast, it certainly is the way to a man's heart through his stomach, they had evidently given up the idea of killing me and when one brought me a mug of stew I know I had been accepted, any hostility seems to have been forgotten, this gave me a good opportunity to study the newcomers, they were dressed similar to Valda there was no difference in the sexes.

But I noticed they didn't greet each the same embarrassing way I was greeted the night before, they all had metal knives and leather pouches hanging from their belts, their tunics were very well made and their shoes were the same as mine, this made me wonder where they got them from, Worra said they had no contact with, them, although their language sounded the same, in fact they didn't look any different from the others on the planet, it's a pity I cant talk to them there's so many things I want to ask if only I had another translator, I'd forgotten about Worra she must be terrified hiding in the cabin while the fighting was going on, and was too frightened to come out, first I had to coax her to open the door to let me, in when I asked her to translate for me she flatly refused to have anything to do with the men, she kept saying they would do naughty things to her, fed up with this excuse i picked her up bodily and carried her outside kicking and screaming, this caused quite a stir, the men crowded around us were amused at the commotion, until Valda ran over and shouted at them to move away, that restored the situation and I carefully put Worra down,, they both started talking and looking at me no doubt calling me a male chauvinistic pig or something. just as bad.

The first questions I asked Worra to translate was to ask where they came from, this puzzled them, when I tried to explain they just said they came from here, and pointed to the surrounding countryside, then I asked why they had attacked me, this must have been an important question as they all tried to answer at the same time, it appeared their ancestors had told stories about the bad man who came with others and built this house, when their ancestors came to look at it, the man pointed a stick at them that made a loud noise and killed one of them, they never came again until today.

When Valda failed to return last night a search party had set out to look for her and tracked her to here, they had just called out to her when I suddenly appeared, the man had rushed at me to stop me getting my terrible stick, the knife had fallen from his belt when I tripped him, then when they saw me hand it back they knew I was friendly, I noticed I was asking all the questions, .they were the same as all the others here, they had their own little world which they enjoyed and didn't seem the slightest bit interested in where I came from, when asked what they do all the time they said they look for water and hunt the dogs, then take them to the food farm where they exchange them for food, clothes, and anything else they needed.

Then they told me something that sent shivers down my spine we also take our old people and change them, I asked what happened to them, but they didn't know or care by their attitude, I asked Worra If the strangers were put here to keep the dog population down, but she didn't know I was disappointed I had been eager to get answers to so many questions, but got nothing.

Valda was quite friendly and even asked to check my dog bites. I was a bit wary of her embarrassing habit on meeting me, and was apprehensive of any goodbye customs they had, but I had no need to worry they just picked up their belongings and left, not even bothering to wave or look back, it still surprises me how indifferent these people are to anything, Worra said I was very lucky I thought she meant about the whole episode and didn't take any notice of her at the time.

The day was too far gone to have another look at the factory, so I finished carving the stock for my gun and managed to fit it together before dark, collecting the rest of my kit ready for an early start in the morning, I told Worra where I was going and showed her how to use my gun, .but to only shoot at dogs, she didn't mind staying on her own it was part of her space pilots training to be independent and she wasn't afraid of the Outlaws any more, although she still wouldn't tell me why she had been so scared of them, in the first place, nor tell me about the nasty things they supposed to do to women, if they manage to capture one, teasing her about this she got quite agitated and went into a sulky mood for the rest of the evening, completely ignoring me at supper time.

Rising very early I was surprised to find Worra lighting the fire and getting the breakfast ready, she seemed to have forgotten her sulks from the night before, I suppose their lack of interest in anything has its good points, breakfast together was normally a quiet affair but this morning she was quite talkative, the atmosphere was different, and I felt I was beginning to enjoy her company, although it wasn't enough to take her with me, she would slow me down too much and I may have to spend the night in a tree out of the reach of the packs of hunting dogs.

Setting off later with my supplies on my back and carrying the old musket, I felt like one of the old explorers setting out for the new world come to think of it that's exactly what I was doing, following the tracks I'd made two days ago I soon caught sight of the factory about four mile away and it would take me over an hour at least, my guess wasn't far out before I was standing looking up at those massive doors again, now I had time to solve the problem of opening them in a rational way, by the look of the damage to them someone had tried brute force and flame cutting to no avail, finding my equipment which I'd hidden two days ago, I sat to think things over, why would anyone build a factory like this make it burglar proof then leave it.

Worra told me the people here didn't build it or they would know how to get in, the actual builders may have left booby traps to stop anyone using them but I think that was very unlikely it would have meant the factories being destroyed and useless when they returned, I believe the stories of the engineers deaths are untrue, they are just warnings to keep people out, Its obvious the best place to hide a key would be on the roof, but its impossible for me to get up there, so lets see if I can make a key, examining the hole wasn't much help, poking my fingers in I could just touch the bottom it ended in a sloping plate, poking a stick in, it seemed like it was just a dead ended hole with no slots at the sides for lugs for keys to fit, all the clues added up to a plain bar fitting in the hole trying the musket barrel it just fitted, that was no good it was too thin it would only have broken it for no reason.

Feeling dejected I thought it could only be opened with a charge of gunpowder which I suddenly realised I've got, that lifted my spirits using the lining out of my pocket I filled it with gunpowder with the intention of ramming it in the hole then realised all the power would fly outwards the only way was to stick the gun barrel in the and fire it, the trouble now I wasn't brave or stupid enough to hold it in position.

Now I thought it over I had an idea, I don't need to pull the trigger, charging the gun with a full charge and leaving out the musket ball I poked the musket tight in the hole; sprinkled some gunpowder along the barrel heaping it over the nipple, now all I had to do was light the gunpowder near the door and retire smartly, bad idea, I lit it and the flame shot along the barrel with lightening speed I didn't have time to turn around before the smoke and sound hit me, waiting for the smoke to clear and the ringing in my ears to go, I was eager to check the results, no good, it looked exactly the same as before, in temper I kicked the door hard, which only ended in pain, bending to rub my foot I saw the burn marks on the door didn't line up with those on the floor, some thing's moved.

Running to the far end of the door I found it had moved two inches exposing an inch gap, now I only wanted something to use as a lever, the musket barrel might be strong enough, finding the gun was easy the plume of smoke rising from the bushes where it had landed led me straight to it, realising the barrel was too thin I used the butt instead inserting it in the gap I levered the door open a bit, then by using flat stones as packing pieces I managed to open them enough for me to get in, expecting it to be pitch dark inside I was pleasantly surprised to find it had a clear glass or plastic roof, the dust raised by the explosion blanketed my vision.

= *CHAPTER EIGHT* =

Standing back to look at the huge doors I was amazed I could move them at all, the crafty lot who had built this place had built the doors in one piece, anyone with average intelligence would expend all their energy trying to part them down the middle, waiting for the fine dust to clear seemed to take ages.

So I turned my attention to the mechanism for opening the door, it looked so simple I could kick myself for not working it out sooner, it consisted of a swash plate behind the hole connected to a floor bolt and a gear meshed to a floor rack that extended the length of the door, all it needed to open the doors was a piece of bar and a hammer, put the bar in the hole, hit it with the hammer, this turned the swash plate lifted the bolt and moved the gear.

Then you came inside inserted the bar in the gear and wound the door right back, there was also a square box welded to the door running just above the rack on the floor this had a conduit running up the door and ended opposite the four square panels or buttons on the outside, they have to be some electrical means of opening, and closing the doors, looking through the haze the sheer size of the place made me feel small, I could imagine a race of giants running the place then after seeing the size of the walkways and gantries. I realised the workforce must have been our size.

Now where was I going to start talk about a kid being let loose in a toy shop I was that exited I didn't know which way to turn first, looking down the main alleyway with branches off at intervals, I decided to investigate each branch as I went along, .the first branch ended in a row of offices, the doors were all open and I wondered in and out of them, this was disappointing there wasn't anything in any of them, no matter where I looked there wasn't any Sign human beings had ever been here, no pictures, no tools no papers, overalls or even rubbish lying about, my impression is this factory was set up as a standby to be brought into production at a later date perhaps, or if their normal factories were destroyed, by wars or earthquake.

With nothing of interest in the offices, I went to examine some of the machines they were all enclosed in glass or plastic shielding and didn't resemble anything I had seen before, two machines had large piles of six foot flat square mesh beside them, it looked like wool with small holes all over but when bent backwards and forwards a few times it resembled metal.

I was standing there in that great mausoleum idly bending the corner and wondering what I could use it for, when suddenly the hair on the back of my head stood up, I was petrified, someone had flushed a toilet, water gurgling down the pan matched the blood rushing to my feet, if my brain hadn't flipped making me incapable of movement, I'd have gone like a bat out of hell never to return, gradually my senses returned to normal my heart stopped trying to get out of my rib cage and my stomach un-knotted itself.

Not daring to breath in case I alarmed who or what ever it was that had pulled the chain, I made my way back to the office where I'd left the musket, loaded it with half charge put a match head over the nipple and waited behind the door straining my ears to pick up the slightest sound, every rustle and squeak magnified a thousand times in my mind, waiting with baited breath I couldn't help feeling a right coward.

I can remember only two occasions when I had been so terrified before in my life, the first time was in the air force, we had just fitted six tubular chairs borrowed from the NAAFI canteen into the back of an Auster ambulance plane, and of course we all piled in for a test flight, leaving the large rear door in the hangar, all went well at first, then the pilot started the test lifting the nose at a steep angle, the seats slipped backwards through the clamps, there was pandemonium as we all slid out of the back of the plane hanging on to anything we could grab hold of for dear life, our terrified shouts alerted the pilot who levelled out and we all clambered back in .ending in a hasty landing and a ground kissing.

I don't believe in ghosts even though I've seen a few, my second scare concerned a ghost, although I prefer to call them hallucinations, I was working late at an old farmhouse at Warndon nailing plastic sheeting inside the windows on the third floor when the same thing happened the hair stood up on the back of my neck and I was terrified, I knew something was standing behind me in the doorway watching me, if it had been the second floor I would have ripped the plastic off, and jumped into the bushes outside, telling myself it was only one of the farm dogs I turned around to see two big green eyes reflecting the light from the windows staring at me I threw, my hammer at them and they disappeared, on retrieving my hammer I found the dogs downstairs. fast asleep.

Thinking about the old days settled my jangled nerves and after a while having heard no more noises I crept towards the next block of offices, carefully pushing the door open I peered inside, it was the toilet block, this is where the flushing noise had come from, but after looking around there was nothing to show who or what had made it, there were no foot prints or marks in the dust on the floor so it wasn't made by humans, looking around at the décor and fittings I had to laugh at them, they were so primitive, the toilet pans were like the old Elsan caravan type, just a tin with a removable liner in it, the water basins were the same and the taps were a work of art the supply pipe came down the wall and bent outwards, a flap over the end with a counter weight served as the handle, being curious to how they worked I pulled the handle over and jumped back in alarm when clear water gushed out, then it shut itself off, this is what had made the flushing noise, they used the taps as relief valves, the tanks on the roof must be quite large going by the amount and the high pressure of water coming out.

Realising it was getting late I had the option of journeying in the dark with the dog danger and sleeping in a big tree, or I could stay here and close the doors, the factory appealed to me although it would be pitch dark; I could light a fire, but it was already getting dark and I didn't fancy going out to collect wood, what I need is oil to make a lamp and there's sure to be oil in the machines in the factory, however I was to be disappointed all the machines had been drained, I suppose if any had been left in the sumps it would have congealed and gone hard.

Now the race was on to find something to burn before it got too dark, luckily when I sat down to think things over I noticed the machine drain taps were about four inches off the floor, and I wondered if that meant some oil was left in them, using a length of railing as a lever to tilt it over, I used one bucket on its side to drain a bit into, then tipped it into the other, looking for something to use as a container for a lamp proved fruitless, in the end I turned one of the buckets over and kicked a dent in the bottom, then filled it with oil, for a wick I used the lining out of my cap rolled up, I also got a bit of the mesh to try and placed it in the centre of the oil, Then I did a silly thing, thinking it would take a lot of heat to get it to burn I turned my fag-lighter up to maximum to light it, both wicks had soaked up a lot of oil and it flared up in my face Brushing off the remains of my eyebrows I became aware of a loud squeak but thought it was to do with the flames, then it increased I looked up to locate the noise when, BANG Whoosh a high pressure spray of water bowled me over, lying on my back I soon saw where it came from, a dome shaped object about two feet across was dripping water, its got to be a fire extinguisher, this really got my attention, If this thing can still work after hundreds of years I could make a fortune with it back on Earth.

THE DOUSER: AN OUT OF THIS WORLD FIRE EXTINGUISHER

While looking at the construction of the ceiling extinguisher I thought why not use a similar portable device for throwing into fires in apartment blocks on Earth, I wasn't harmed when they went off here, so they wouldn't hurt anyone left in their room, it would be very useful in high rise apartments.

Just a strong plastic bottle full of water with a crow-scarer type of firework inside it, stick a blob of phosphor for a fuse on top of the firework and arrange a strip of emery paper to scrape it when pulled, this will ignite it, setting off the firework to pressurise the container, the bottle will need to be perforated with small slits then painted inside with a plastic paint to make it water tight this will rupture under pressure. letting the water squirt out.

An accordion folded metal tube around the firework lets it expand without splitting and will stop the expansive pressure exiting in one place, coating it with plastic will seal the ends and protect it from corrosion.

Sat thinking about the various uses a device like this would be useful for I conjured up visions of squads of firemen turning up at an American skyscraper fire, realising their hoses were useless sending for the hit squad who arrive with a small howitzer which they load wit these gadgets I'll call Dousers for now, aim them at the windows and fire them by compressed air.

The Dousers would smash the window enter the apartment then expand showering the place with water delaying the fire long enough for the firemen to enter safely.

I had a little laugh to myself when I thought it may be better to use old fashioned gunpowder as that would burn slower giving a more controlled thrust to the water ejection, in effect using eighteenth century Earth know how coupled with alien design.

I had noticed the domes fitted in the offices, but thought they were lights, there were no switches on the walls so I hadn't bothered with them.

The extinguisher device was very efficient, everything was soaked including me, emptying the oil and replacing the wicks with mesh gave me a good light, now I had something to interest me my tiredness disappeared, the device had gone off so it would be safe to examine it and see how it worked, the problem was how to get it down, I'd already collected a pile of mesh to make my bed so I had enough time before dark to find something to stand on, I soon found a length of railing which dropped into slots on a machine, dragging it into the toilets, I wedged the bottom against one wall and propped the top against the opposite wall it fitted perfect, in fact I wondered if the original builders used the same method to replace spent units.

Climbing my makeshift ladder I had a bout of coughing brought on by the smoke hanging in the air, this almost made me give up the attempt and leave it till morning, but now I've gone this far I may as well finish it, examining the device I found it was made of shiny metal like aluminium it had rows of fine slits all over which looked like they had opened up under high pressure allowing about three gallons of water to squirt out, in a few seconds it had swamped the room and put my fire out, normally water on oil fires spreads the flames rather than smother them.

This idea would be great in apartments, or even to replace sprinkler systems, perhaps it could be used for portable extinguishers to be thrown into fires by hand or machine, another good point it can be used where people are present.

The device going off didn't harm me, even though I was looking up at the time, Getting back to the job in hand I grasped the side of the rim to try and get it off, wrenching the whole thing each way I felt it move expecting it to be a screw fitting I hung on trying to unscrew it, I was caught unawares when it suddenly came away and to stop myself falling I had to let go, it had been attached to the ceiling by a bayonet type fastening, it bounced off the ladder and clattered to the floor in pieces making enough noise to awaken the dead, feeling a bit stupid I dragged the ladder outside, and returned to collect the bits and examine my prize in detail.

Fitting it back together, I saw the simple way it was made starting from the outside, first a metal dome shaped cover with a wide lip and pierced all over with slits, above this was a ridged inner plate lipped over the edge, then came the main backing plate reinforcing the edges and holding it all together was a band around the outside, by the look of the burn marks the explosive charge was in between the two lids, when it went off it deformed the thin ridged plate forcing the water to open and exit the slits, finding out what made the screech noise was beyond me, but the sensor and igniter was a work of art, this was attached to the bottom centre and looked like a four inch figure of eight with a gap in the middle this evidently clamped on to a flat spring, when it warmed up it expanded and released the spring which swung and hit a firing pin sited in the middle of the device.

Feeling very pleased with myself at my discoveries, I now turned my attention to boiling water, I didn't fancy drinking water that could be hundreds of years old, taking one of the toilet outer bins I bent four sides in a bit, placed the deformed part of the extinguisher filled with oil and two wicks under it then put a new toilet on top, ...Hey Presto, my new boiler, lighting it with more caution this time I soon had two gallon of hot water boiling away, the room was hot and I was soaking wet so it was a good opportunity to wash my clothes, the way they felt sticky there could have been chemicals in the water, besides it could have been stagnant in the extinguishers for hundreds of years, having no soap was a pity but I stripped off wedged the door and gave my clothes a good washing. By using the grape treading method.

While the water was boiling I tore some mesh into strips and made a clothes line tying it between two taps and throwing my clothes over it to dry, now I threw away the water and boiled a second lot filling my water bottle and leaving it to cool, in a sink full of cold water It was beginning to feel like a sauna in the room so it was quite pleasant to have a strip down wash, my shirt had nearly dried so I used it for a towel then washed it again checking the oil level in the fire I topped it up, made sure the door was wedged open enough to let the smoke out then I lay on my bed of mesh thinking about the days events and before I knew it I had drifted off to sleep.

Waking up starving hungry and having nothing to eat wasn't a very good start to the day, so far I'd done very well, and now I was going to add to my good fortune, getting my makeshift ladder I placed it in the second office under the fire extinguisher, then used my fag-lighter on maximum to heat the sensor, the Squeak started and I managed to get out before the deluge, It only took a matter of minutes to take it down and dismantle it, the parts would come in handy back at the cabin

Clearing up and putting everything in order in case I had to spend another night in here, I started off to explore the rest of the place, the big dome in the centre interested me, approaching it at ground level there didn't seem anything unusual about it, but when I climbed over the mass of pipes that surrounded a large deep hole in the floor it took my breath away, the pipes were connected to a large outer ring about fifteen feet across and an inner ring level with it, was suspended from the roof by more pipes, I saw what the mesh panels was used for, half a Zeppelin shape spaceship was growing up from the bottom of the pit and some panels were already in position, they were butted together and the mesh was put in place then sprayed or electroplated.

Climbing down into the hole I could only marvel at the simplicity of it all, on Earth our technology is getting more technical and costly, here they have gone the other way perhaps they did the same as we did in the past using up all their natural resources in wars and are now paying for it by having to skimp and save.

Scratching the surface of the plating didn't leave any marks nor were there any burn marks, this ruled out welding, so it had to be a form of electro plating, although the size of the cables or rather pipes going to the rings were only about half an inch, diameter the only conclusion I drew from this is that the current is a high frequency rather than high power, the pipes to the outside rings are red and the inside rings are blue.

I suppose that means the current flows one way, another unusual thing was the thickness of the metal on the finished part was about an inch thick, it looked like the mesh matting swelled in contact with the sprayed metal before it hardened and made a porous very strong shell, the outside was smooth and shiny, looking down the inside it was the same, with built in deck buttresses running straight down, opposite each other, I imagined what it would look like when the whole machine was working, the rings would expand and contract to make the cigar shape, then I expect the rings would stay at the top while the shell was lowered down onto trolleys and wheeled out.

Turning round to climb out my pocket caught on a injector and out popped my fag-lighter to go clattering down into the depths of the pit, I'd be lost without my precious lighter so I had to go down to find it, searching around at the bottom I had a good look around, something glinting on the floor caught my eye, excited that at last I had found something interesting, I clambered over the pipes, and picked up a pair of sunglasses, what a let down, they were as much use to me as a bloody chocolate tea pot.

= CHAPTER NINE =

Checking them over after getting out of the pit, they were just ordinary welding glasses, there was no name to identify the makers they must have belonged to the engineers who tested the machines, this proved they were humanoid and the process to make the shell involved very bright light, the lenses were so dark I couldn't see my fag lighter flame through them. carrying on down the factory, not stopping to look at anything else I arrived at the back doors, I found they were exactly the same as the front ones and the bar was still inserted in the gear wheel, winding the door open enough for me to get through was quite a struggle, the front door had been much easier, perhaps the sudden jolt from the explosion helped to free it. Squeezing through the gap the bright sunlight dazzled me for a while, what a surprise I got when I finally managed to look, I don't know what I expected, to see but I know it wasn't the sight that greeted me.

It was a scrap yard, not any old scrap yard, this was a scrap yard of spaceships I counted four complete shells and six wrecks with most parts missing, it would have taken an army to have created so much destruction, I don't think the missing bits were used to make new ships, because the missing parts were mostly the same shape. I got the impression they were used for repairing or making something else, climbing in and out of the wrecked ships I found sections with doors on, which I thought if I fastened them together I could make myself a shed of some sort to sleep in.

Worra won't worry if I don’t return to the cabin to sleep tonight, none of her race seem to worry about anything, this time their indifference will work to my advantage, soon I came to one of the untouched ships it was the size of a barn, walking around it i found it was similar to the one that brought me here although it was much bigger, the door on the far side had a few dents in it, but I don’t think it had been opened in a very long time, in fact the way the trees had grown up through the wrecks showed nothing had been disturbed for a few hundred years.

The door had the same four black panels at the side and a hole near them, this is going to be easy I thought smirking to myself, just get the old trusty musket load it poke it in the hole and Hey Presto Open Sesame, that was in theory, in practise it didn’t quite work out that way, I loaded the gun put a match head in position inserted muzzle in the hole and pulled the trigger, the powder in the pan spluttered, And that’s the last thing I remember until waking up with a splitting headache and a heavy weight pinning me to the ground, panicking I pushed the weight off me and tried to get up only to fall over again.

I was too dazed then as my head cleared I realised what had happened, there had been a build up of gas in the ship and the sudden opening of the door had released the pressure blowing the door on me and I had a bloody nose to show for it, the stench from the open door was appalling there was no way I could go in until it cleared, while waiting for my nose to stop bleeding, I made four snares from the mesh and set them up a short distance away collecting some water at the same time served its purpose the bleeding had stopped and the stink wasn’t so bad.

In my eagerness to explore the ship I jumped in through the doorway, the smell made me wretch I couldn’t stand it and jumped out again, this isn’t any good I’ve got to get rid of the smell and get a light of some sort, it was pitch black in there, sitting outside thinking how to get rid of the smell I noticed the wind had increased, it usually did this time of the day to blow the morning mist away, if it can be diverted into the ship it will solve my problem, lifting the door was easy, things don't weigh as much here as on Earth, the door being bigger than the doorway it was easy to jam it half way across the opening, standing down wind of the door I could feel the draught coming out of the ship bringing the over powering smell with it.

Now things were looking better the next thing I wanted was a lamp, looking amongst the scrap it took me a long time to find a suitable dish shaped piece of metal, pouring some of the gearbox oil into the dish and dipping in a piece of mesh for a wick I lit it and ventured into the unknown, getting used to the smell didn't make it any easier, but I was so keen to find out what was inside the ship I would have braved anything, the blue flame from the lamp made everything look a dirty grey colour, the first things I saw was a stack of boxes about three foot square with hinged lids, which I couldn't open as they were stacked on top of each other right up to the roof, and touching both sides, a solid wall of boxes.

By the side some long boxes were tied to the floor with those easy to undo straps, eager to look in them I put the lamp on top and walked around to get at the farthest straps when my feet got tangled up in something on the floor and I only managed to save myself by grabbing hold of the lid, leaning over I used the lamp to light up the object I'd fallen over, a pair of boots attached to a pair of legs came into view, moving the lamp the rest of the body came into sight, and what a sight, it resembled one of those ancient Egyptian mummies lying there on its back, it was a horrible gruesome looking thing to look at, but I was too curios to be squeamish, holding the lamp closer for a better look, the body wore a blue overall with a belt around the middle, the only things of interest were some objects attached to the belt.

Feeling guilty about desecrating the dead I pulled the belt apart and rolled the body over to get the belt free, a cloud of dust arose a horrible stench followed and of course the bloody head fell off and rolled away into the darkness, this was the last straw feeling sick I dashed for the door and only just made it in time, before being violently sick, this and the fresh air soon revived me, although now I felt really starving hungry as if I hadn't ;eaten anything for a week.

However I soon forgot my discomfort when I checked my prizes. the belt had three oblong flat cases that looked like the remote controls for opening garage doors, two were two inches by three inches and the other was two by four, each had a glass square and four small black squares under it, the two were identical except that the black squares on one were worn, as if it had been used a lot, and the other was brand new, leaving them on a flat rock hoping to study them later, I set off to check my rabbit traps, moving as quietly as possible, going from trap to trap I was suddenly surprised to hear voices, creeping up to the source I found a party of four guards taking a break, evidently they must be looking for me, if I'd been outside the ship they'd have found me.

The only time they want me is when one of the bosses from the home planet wants to interrogate me, lying there watching them eating made me feel starving, I hadn't eaten since last night and now it seemed it would be a long time before my next meal the leader who had been looking at a small hand held device shouted out a command and the four of them picked up their kit and moved off at a fast pace.

Waiting until they had disappeared out of sight and making sure they were going to pass the ship on the opposite side to the open door I went to see if they had left anything, one of their metal bottles had fallen over and was still half full this and a couple of biscuits they had left behind helped a little, then I realised if they return this way they will know I'm around by finding the bottle empty, the only solution was to relieve myself in it, they may not notice the difference unless they drink it, .the thought made me laugh, it would give them something to puzzle over for a change.

Getting back to the ship and seeing a wisp of smoke coming out of the open door made me realise how lucky I'd been, if the search party had seen it and investigated they would have found my lamp still burning inside the ship, perhaps the smell of the gas had put them off, they would also have found the three control gadgets I'd left on the rock by the door, picking up the worn one and pressing all the black squares on it didn't do anything, picking up the odd one I pressed the first square, CLANG I leapt in the air in fright, I whirled round expecting to see someone who had crept up on me.

After checking around to make sure there was no one, then looking to see if anything had fallen down, I carried on playing with the little box, pressing the same square again did nothing but when the second square was pressed CLANG. there was that noise again, and it came from the door, sat in the door opening, frightened to press the buttons on the boxes in my hand in case it was my imagination, I couldn't believe it, this device must be hundreds of years old and yet it still worked, pressing them alternatively in rapid succession just to make sure it was the bolt on the door opening and closing, showed it was working, there's no rust anywhere but the vegetation engulfing the wrecks means they've been here a very long time.

Thinking about time reminded me, what am I going to do with the skeleton and the head, it's a job I didn't fancy doing, but there's no one else so its up to me, fitting a strip of cloth over my face picking up the roll of mesh and some more oil I didn't want the lamp going out on me while wrestling with a skeleton, telling myself its only a bag of bones its not going to attack me, I went in filled the dish with oil then turned my attention to the bones, the clothes were still in good condition, as the head was missing it was easier to drag it outside by pulling on the trouser legs, finding the head was different, searching behind the crates It had wedged itself in a gap picking it up by the hair it and half the face came away in my hand, now I got mad at myself for being so squeamish, I hooked my foot around it dragged it out, then kicked it down to the doorway and outside, kicking it and, dragging the body a short distance away from the ship, I piled some stones and metal sheets over it said a prayer stuck a cross over it and returned to the ship, while looking for the head I had seen some small boxes at the back and wondered what was in them, undoing the straps and dragging them to the door the lids were easy to open and I eagerly sorted the contents this must have been a personal tool box there was a pair of overalls, boots some towels which fell to bits, and a writing slate, the boots were too small but the overalls fitted me.

Outside the sun had gone down and I prepared to spend the night in the ship the smell had cleared a lot and with my roll of mesh it wouldn't be so bad, putting the door back in place I settled down but was too excited to sleep, the noise of the wind on the hull magnified by the size of the place and the creaks and groans made it feel proper creepy, had I known what was on the upper deck I wouldn't even be in here, I would have braved the packs of dogs and slept up a tree, I eventually dropped off.

Only to be awakened by a fit of coughing a splitting headache and the vile smell had returned, the lamp was still burning so the air was breathable I don't know what would have happened if I had woken up with no light, imagine stumbling around in the pitch dark trying to find the door, in future there will be two lamps burning, staggering to the door, I worked the remote control and kicked it open It was a hard job to wedge it across the doorway to divert the air into the ship again, but it soon cleared the smell, the fresh air also cleared my headache although my sickness stayed with me all day, having nothing else to eat didn't help.

Using the remote controls made me think, what was the other black squares for, the only way was to go around the ship and try them, walking around and trying the buttons didn't produce any results only disappointment, it appeared there's no way to get to the upper floors this was getting too silly for words, there has to be access of some sort, having done a complete circuit of the cargo bay and finding nothing, I almost gave up hope, the stink was still there, so I stood in the doorway taking deep breaths of fresh air and fingering the buttons, I heard a click thinking it was the door clamp I took no notice, even when a faint shadow appeared on the doorpost It failed to register, it was only when I jumped down and looked back the whole of the bay was getting lighter, a long tube all around the bay was lighting up gradually until it was as light as day, how the hell could they store power for so long the only solution I could think of was that they used solar panels.

Having got the doors open and the lights working lifted my spirits and made me keener than ever to get to the other decks, the first deck only held cargo and that was only in large plastic bins they in turn went right up to the ceiling, it looked like they were put in place by pulleys and cables, until I could get the power working I couldn't open the lids, however there were some more long boxes secured to the floor by straps these were easy to open, lifting the lid cautiously I peered in, Christ, its another body, this spaceship must be an undertakers hearse, either to bring bodies here for the food growers or to take radio active ones to dump in space, dropping the lid quickly to cover my gruesome discovery, I sat for a while to wonder why there was no horrible smell.

My curiosity overcame my fear and lifting the lid again I reached in and squeezed the leg there were no bones in it, neither was there a body, bringing the lamp closer I could see it was a space suit with a space mask in position, it even had the flexible pipe connected to two air bottles held in a cartridge type holster on the belt, no badges or motives adorned the suit, and there was nothing in the pockets, in fact it looked brand new, packed around the suit were about twenty spare air cylinders for use with the suits, this was unusual they don't normally overdo things, it's the first time I've seen any extra spare parts attached to anything.

= CHAPTER TEN =

The suits were too lightweight for use in space, and the openings down the front were not airtight, it had those push together fasteners, I think they were for use in the mines in the presence of gas, noticing the oil lamp was getting low and not knowing how long the main lights would last I decided to look for access to the other floors, now I had better light I soon found another door I had missed on my previous search, it was fitted so well I nearly missed it the second time, the remote didn't work until I held it right up close to the panel and worked it a few times.

I could see the sense in this otherwise it would have opened all the doors in the ship, a little puff of air went past me as the door slid open I didn't take any notice at the time, the room was empty but quite large, crossing the room to examine a small panel on the wall I felt the floor move of course it's a bloody lift, I cant think why it took me so long to find that out, it's the obvious thing to have between floors, at this point my courage ran out, would anyone trust a lift that hadn't worked for a hundred years or more,, what if it got stuck between floors, what if the doors jammed, what if the next floor doors didn't open, all these problems multiplied in my mind a thousand times.

Leaving the lift I went outside to think things over, I decided now I've come this far go for it, there are sure to be access panels, collecting my kit I made for the lift on the way I passed the long boxes, laughing at my stupidity in thinking the spacesuits were bodies, I realised a spacesuit was what I really needed, putting on the spacesuit was easy it was very light weight and made of very thin plastic that stretched like rubber, it must be a one size fits all, the belt had three pouches on it, which meant it held two spare cylinders so they must last a long time, checking how to fit a spare cylinder I found it difficult, it wasn't until I lost most of the pressure that I realised the pressure kept them in position, and it wasn't until they were nearly empty that they could be changed, that's why there were no gauges, the face mask fitted over the head and fastened to the suit by the now familiar Teasel type hooks.

Having the suit and air supply put most of my fears at rest, the next problem was light, the lamps may not work on the next deck and it would be pitch black up there, Carrying my oil lamp with its naked flame didn't appeal to me, although the miners in the eighteenth century coal mines used them that way, until a chap named Davy put a mesh around the flame and made it safer, going back to my oil lamp I put two layers of mesh around it to be on the safe side, it cut the amount of light down a lot but I thought it was better to be safe than sorry, it still worked burning with a light blue flame, placing it in the corner of the lift I turned my attention to the controls.

Turning off the lights in the cargo bay to conserve energy, I noticed the lift panel gave off a dull glow that was a good sign it showed the power was on, using the remote to close the door I concentrated on the wall panel, touching the top square resulted in a loud clicking, pressing the others in turn didn't give any response, talk about disappointment, after feeling so jubilant at getting this far, to be let down at the last fence really upset me.

Thumping the panel in frustration resulted in a loud clicking, I waited for a few minutes to see If anything else was going to happen then gave it another thump, this only stopped the clicking noise, Now doubts were beginning to plague me, just because the door opening and lighting controls worked it didn't mean all the other controls would, thumping the panel again just to check, a chattering noise started then abruptly stopped, to be replaced by a whirring, then without warning the lift jumped and stopped, O Christ the lifts stuck, my worst fears were justified, then the door opened, WHOOSH a burst of pressure flung me against the back wall knocking all the breath out of me, I lay there terrified, watching the flames from the knocked over lamp flare up.

Fully expecting an explosion and not being able to do anything about it, I just accepted I was going to die, after what seemed an eternity I opened my eyes I couldn't believe it, the flames where dying away the build up of gas in here wasn't inflammable, it also meant I wouldn't be able to use my lamp, there wasn't enough oxygen to keep it alight, then I noticed the place was getting brighter the lights were gradually coming on, they must come on automatically when the lift stops.

This really got me wondering surely they cant be working off batteries that could be four or five hundred years old, another thought struck, me where was the power coming from to work the lift, the spaceship that brought me here had some sort of engine, because it started up when we left Earth I heard it working when we changed direction, and again when we landed here, but I haven't heard anything working on this ship, in fact its deathly quiet, just a few creaks and groans as the metal shell expands and contracts in the sunlight.

Thinking about the deathly silence reminded me of the body I tripped over on the lower deck, surely he wasn't the only crew member, there wasn't any more on the lower decks otherwise I would have found them when looking for access to the upper decks, now its dawned on me, the burst of pressure that knocked me over when the lift stopped and the door opened, can only come from decomposing bodies and I've a feeling there are lots of, them, but where can they be hiding.

The breeze created by the door opening swirled around the deck carrying clouds of fine dust making it hard to see, and making me appreciate the spacesuit and air supply I was wearing, waiting for the dust to settle, I eagerly went to examine the row of cabinets that stretched the length of the ship, the first ten lockers on each side had bodies standing in them, the others were empty, I tried to open the doors but as it later turned out later I'm glad I didn't. I had a horrible feeling the bodies were alive, perhaps they were in suspended hibernation or something and could start to regenerate, now I've switched the lights on, banging on the doors to dislodge the dust, then straining my eyes only resulted in shock, a large bloated face with bulging lifeless eyes stared back at me, this horrible sight coupled with the deathly silence of the ship put paid to any more exploring on this deck, I hadn't found anything of interest, this puzzled me, there had to be a control centre of some sort but where, the lift had only two panels to press and it seemed obvious one for up and one for down. I'd only pressed it once.

Perhaps it has to be held down, I couldn't get back to the lift fast enough to try it out, sure enough one press and up I went, and the door opened, expecting a blast of air I held on to the lamp, but nothing happened the pressure being the same up here meant there's more body's And there were, as the lights came on automatically the dust cleared and I saw three bodies lying just outside the lift, they didn't look a very pretty sight either, lying in a pool of water they were blown up like balloons, it was only their space suits that stopped them bursting. Lying on their backs the bulging eyes reflecting the overhead lights appeared to follow my every move, one had a face mask on and the others had their mouths wide open as if gasping for air.

Looking around for clues as to how they died I could only guess, by their positions it looked like they had been sharing the same air supply, there was only one air bottle so why didn't they go for the spares, or did they send one chap for them and he fell or was overcome by gas, a race that could build spaceships would have fitted emergency systems into it, perhaps the life support system had failed, feeling I wasn't getting anywhere I had a quick look around, but there was nothing of interest for me.

The layout of the control console was similar to the one on the ship that brought me here, this gave me an idea, Worra would know all about the ship and how it works, I can bring her here and get the systems working, besides I've hardly eaten in two days, its only the excitement of exploring this ship that has kept me going, now I've decided to bring her she will want a face mask, I wasn't sure if there were many down below, gritting my teeth I went over to the bodies and pulled the mask off the one wearing it, this was easier than I expected, not feeling so squeamish now I tried to go through their pockets, but the clothes were pulled too tight.

They all had the same uniform with no badges of rank or motifs of any sort, the only things of any interest were their belts and the objects on them, they were the remote controls for operating the lights and doors, slicing the belt off one body I pulled it away and found a different type of control, this one was shaped like a letter T made with a line of balls it looked like a large key I put this in my pocket to examine later.

Collecting my kit and all the remote controls I put them in the lift had a last look around to check, when really there was no need to there was nothing to leave behind, going down in the lift I began to wonder why did the people who built this ship make everything so basic, there was no decoration or paint anywhere, no creature comforts, no pictures, or personal possessions, even the padding on the seats was blown up plastic.

Reaching the ground floor I was relieved to find the lights come on, they must be connected to the lifts whenever I use them the lights work, my lamp which I'd left burning had gone out which was just as well, the smoke drifting out of the door could have attracted attention, which reminded me, if I'm caught with the remote controllers on me they'll know I've gained access to the ship, it could be to my advantage to keep this from them, putting the door back in place, and locking it, I put the controllers on top of a high piece of wreckage, that way they were well hidden and would be fully charged when we needed them.

I kept one of the remotes in case it worked the factory doors, otherwise it meant I wouldn't be able to lock it behind me, then went to check my snares which I'd set earlier, two of them contained rabbits and it wasn't long before I had their coats off and cooking them over a blazing fire, I found some water in pools in the wreckage there were no wriggly things in it, but I boiled it to be on the safe side, although I did burn my tongue in eagerness for a drink, after a hearty meal, I know it doesn't sound very appetising burnt rabbit and hot water, but you have to be starving to appreciate a meal like that all I wanted now was to sleep, that was out of the question I only had enough time to get back to the cabin before dark, no way am I sleeping out tonight, the smell and dead bodies are having an effect on me in daylight so what will it be like in the dark.

Busy stamping out the fire I only noticed the men surrounding me when I bumped into one as I turned around, and it was too late to do anything, raising my hands in the air seemed to surprise them, the leader took my bag off me and gave it to one of the men then pushed me in the direction he wanted me to go, noticing he had a translator fitted I switched mine on and asked where he was taking me but he didn't answer, my meal had got rid of my hunger pains but I was still dead tired and wasn't looking forward to a long walk back to the interrogation centre as I'm sure the reason they want me is to answer more questions.

Arriving at the front of the factory I was relieved to see one of their three wheeled buggies and trailer waiting, I was motioned to sit on the back end of the trailer which I thought was a funny place to sit a prisoner, while the rest were sorting themselves out, one of them decided to have a drink from his water bottle, well it couldn't have worked out better if it had been rehearsed, he must have picked up the metal bottle I relieved myself in and being in a hurry to drink before the car moved off he took a big swig, the performance he gave on finding it wasn't water would have won him acting awards, he gave out terrible screams, gurgling and wailing at the same time coupled with such fantastic gyrations raising the dust it was a job to see him at times, the commotion eventually ceased as he collapsed in a heap, I don't think they're used to such exertions he may have died as far as I know, while they loaded him on the trailer I could have escaped quite easily instead, I found the opportunity to get the remote out of my sock scoop a hole near the wall and bury it.

I couldn't do much for laughing anyway, I often wondered if they knew what or who caused such a brilliant performance, I know I was the only one to appreciate it, our overloaded buggy trundled along in silence, we finally arrived at the interrogation centre. My escort couldn't get rid of me quick enough pulling me off the trailer and pushing me into the showers with my clothes on, then they signed for me to take my clothes off, I thought it might be a punishment for playing the bottle trick on them then it dawned on me, I must have stunk to high heaven, having been inside the spaceship for such a long time with those rotting bodies the smell would have penetrated all my clothes, no wonder the chaps on the trailer had kept away from me, and sat me on the back. They soon brought me a new set of clothes and urged me to hurry, the person I'm about to see must be very important, it's the first time I've seen concern on their faces, taking my time dressing, just to watch them get more agitated I couldn't drag it out any longer and set off with two of them as my escort.

= CHAPTER ELEVEN =

The tables had been arranged in a half circle facing me, with a guard on each side I had no chance of escape, a newcomer had joined the original team and she captured my attention straight away, she was dressed differently with a roll neck jersey and a belt, I imagined her with a pair of jackboots, She was also the only one wearing a translator, which struck me as very odd, if they couldn't understand what was being said why have them here, the way they treated her she must be high on the social scale, in, fact with her bearing and aloof attitude she looked every inch a Duchess, she was in charge of the meeting and soon let everyone know, starting the proceedings by snapping a question at me.

"Why did you turn your tracking signal off for long periods of time, we had great difficulty finding you."

The question caught me unawares it hadn't occurred to me the translators gave out long range signals, I didn't want them to know of my success in entering the factory or ship, so I hesitated before answering, without any warning she picked up a key shaped object from the table, similar to the one I found on the ship, and pointed it at me a bright red flash settled on my hand, the next thing I remembered was picking myself off the floor and painfully getting back to my feet, , "Next time it will hurt more."

The voice broke into my fuddled brain as the progress of events came back at the thought of another shock.

"I didn't turn it off I squeaked" my neck muscles hadn't fully returned to normal from the effects of the electric shock, Before she could answer she was interrupted by one of the shower attendants who gave her a roll of cloth, she unrolled it and I saw it was the space suit I had worn in the space ship, she suddenly flung it at me screaming.

"Where did you get this."

Then she grabbed for the gun again, well I wasn't going to stand still for target practice, I scooped the suit off the table and flung it back at her, she lifted her hands to ward it off just as she fired, the beam caught the guard on my left full in the face he grunted and collapsed in a heap the other guard turned, I caught his arm and pulled him in front of me her next shot must have hit him in the back he yelped and fell on top of his companion not waiting for her to aim again, I leaned across the table grabbed her arm with the gun in it, and pulled her bodily over to my side. of the table.

She and everyone else in the room were in a state of shock, taking the gun from her outstretched hand I turned to the frightened onlookers waved the gun at them and said those time honoured, words, "Anyone who follows gets it" They must have understood because they sat still and put their hands on their heads, now I had time to think, Christ what have I done this time, now I'm really in trouble, the two guards were still unconscious on the floor, I could escape quite easily but they would only come after me, then id have to start shooting them with this funny gun.

"Do something "the Duchess screamed at no one in particular, this startled me into action, grabbing her arm I pushed her through the door. "Just keep walking, do as your told and you wont get hurt", this must have been said in every gangster film ever made and even in my tight situation I had to smile at myself for saying it, luckily the car was still outside where we'd left it,. "Get in and drive "I ordered her as I wanted to deal with any pursuit" "I cant drive" she wailed but that wasn't going to stop me.

Picking her up I dumped her in the driving seat and said, "press the pedal," we shot off at full speed I let her carry on at this hectic pace as it would give me time to work something out and organise my defences at the cabin.

Trying to keep a lookout and grabbing the handlebars now and again to keep us going in the right direction, we soon came to the cabin, and went straight past, she didn't know how to stop, we carried on at speed bouncing up and down until I caught hold of her around the waist and lifted her bodily out of the drivers seat lifting her foot off the pedal automatically slammed the brakes on.

All three wheels squealed and locked on, we skidded about twenty yards then did a somersault throwing us both out, I couldn't see anything of the Duchess until the dust settled, then I saw her a little way off lying motionless," Bloody hell" if I thought I was in trouble before, what trouble will I be in if she's dead, how can I explain it, thinking of all the dire consequences that would happen to me I approached the body with dread, she moved and sat up, my spirit soared, then the most amazing thing happened she burst out laughing, I couldn't believe it, they could actually laugh, I've had lots of surprises since coming here but this one beats them all, tears were running down her cheeks washing little rivulets in the dust down her face then dropping off, I was so exhilarated and relieved at her recovery I picked her up and kissed her, she squealed and fainted, "Christ I've never had that happen with a girl before," my technique must have improved a lot.

At a loss what to do now, I went to look at the damage to the car it had ended up on its side, rocking it back and forward a few times I soon got it back on its wheel driving in a circle I stopped to pick up the Duchess then carried on to stop at the cabin, Worra was waiting by the door she had seen us fly past and come to an inglorious stop in a cloud of dust and broken bushes, but she didn't say a word, I wasn't sure if she was sulking or if it was her natural indifference, I had told her I would only be away one night, then she caught sight of my passenger stared in disbelief then turned and ran inside the cabin, before I could do anything she had locked and bolted the door," Bloody hell" what have I let myself in for now. I've got two women to sort out.

I'm tired and hungry and I've got to get in to defend the place before a rescue party arrives, the cabin is built like a fortress I cant break in, trying to talk her into opening the door was a waste of time, I even thought of ramming the door with the car although it gave me an idea it was a gamble but these were desperate times.

Driving the car to the side of the cabin I rocked it back and forth and tipped it against the side of the building, then used it as a ladder to get on the roof, smoke was coming from the chimney and it didn't take long to block it completely In a surprisingly short time I heard the bolts being drawn the door flung open and Worra came staggering out coughing and wheezing her eyes streaming with tears, I felt really sorry for her but as I said desperate times call for desperate measures, shoving her in the direction of the Duchess I climbed back on the roof to remove the blockage and look for any signs of pursuit, nothing appeared on the horizon in any direction.

Apparently they hadn't organised a rescue party or perhaps they hadn't got any more cars to travel on and would have to get here by walking, in that case I'd have about two hours to organise my defences. Jumping down in haste in case Worra had recovered enough to run back inside and bolt the door again, I found her being comforted by the Duchess who was patting Worra on the back, the smoke had really affected her, this made me feel guilty as hell for causing her so much discomfort, if I told her Earthmen rolled dried leaves in tubes of paper put it in their mouths, set fire to it, then sucked in the smoke for pleasure, I don't think she would have believed me.

Leaving the two women outside I went to check the defences the place was a proper fortress Rip Van had built it to withstand Indian attacks, although I had no idea what any rescue party would do, their stun guns were very effective at short range, that reminded me how are they fired there's no trigger on them, taking the tee gun from my pocket I pointed it at the wall and tried to fire it, try as I might nothing happened in the end swallowing my pride I went outside and asked Worra, she had fully recovered by now and showing the usual indifference to the fact I had nearly suffocated her by showing me how to shoot.

The first squeeze on the ring part produced the thin red beam for aiming, further pressure made it click and flash yellow, it feels like it works by a pietzo crystal,. guns like this would be invaluable to police forces for questioning suspects, the Duchess had been talking to Worra and she must have said a few good words about me, as she accepted her role of hostage without question, when I asked them about the gun they both said it never needed charging or reloading, using it repeatedly would cause it to overheat and ruin it, not knowing how long it would last left me in doubt, I couldn't shoot anyone with my rifle or musket, I reckon I had seen too many cowboy films to even think I could fight them off, and as for killing them I knew I wouldn't have the nerve, having time to consider I think my best course is to take the Duchess back and hope she would put in a good word for me.

Sitting down with them both eating the last of the biscuits I asked the Duchess if she would cooperate and get me out of the mess I was in as a lot of people could get killed, and that was the last thing I wanted, she looked thoughtful for a moment then switched her translator off, and had a long talk with Worra in their own language, finally switching her translator on she said she would do as I asked on condition I showed her how to enter the factory, she knew I'd been inside by the different coloured space suit I was wearing when caught, of course I agreed it was no use denying it, in fact it will suit me fine, if we can get some of the machines working Id be able to pinch their ideas. Seeing a look of interest on her face I thought now was the time to ask for favours.

We had run out of supplies and it would be a long way to get more, she asked what things I wanted, I said it was just food.

Now I got another surprise, she touched her translator at the back and spoke in her language for a few minutes then said they will be here in the morning, Bloody Hell it's a translator a tracking device, and a transmitter I'd love to know how they work, they must be powered from being charged from sunlight that's why we have to switch them off at night, she must have been laughing at me all the time when I thought I was in control, no wonder she was their top interrogator, I was that niggled I didn't offer her my bed for the night, Worra said the Duchess could use hers, I was surprised to hear this, the people here don't usually give or lend anything, Worra settled for the armchair and I think she was the first to fall asleep.

A loud yelp woke me from a beautiful sleep I lay still for a few minutes to collect my wit's then realised something was wrong looking around I saw the Duchess in the doorway pointing the tee gun at something outside, trying to remember what had happened the night before and dressing in a hurry with my befuddled brain didn't go together very well, I overbalanced and crashed to the floor in a tangled heap of trousers and bedclothes.

The Duchess whirled around which put paid to my plan to rush her and get the gun, standing with my trousers around my ankles didn't embarrass her at all, she came over and threw the gun on the bed, I shouted to ask what the noise was but as I hadn't switched my translator on she didn't understand me, running outside to find out, what was happening, the sunlight dazzled me and I fell full length over a body, on the floor, getting to my feet and dreading to look round I was relieved to find it was one of the wild dogs lying there, now was a good chance to have a look at one, although it wasn't a very good specimen, I suppose I'd look thin and weedy if I only lived on rabbits.

= *CHAPTER TWELVE* =

By the time I'd collected my kit together a dust cloud appeared on the horizon and a three wheeled buggy soon lurched to a stop beside us, the two occupants got out and went to look at the car tipped up against the side of the house, I told them it was broken which they didn't understand, the Duchess confirmed this and they didn't bother to look any more they took her word for it, we unloaded half the supplies in the cabin and left half on the trailer to take with us, leaving Worra to look after the place, we boarded the car and trailer the Duchess surprised me by ordering the driver onto the trailer then jumping in the drivers seat herself. And moving off.

Remembering her previous driving skills I hastily instructed her in the art of stopping by stamping on her foot and letting it come up slowly this brought the car to a stop, starting off at a slower speed we had a more pleasant ride and in no time at all we arrived at the factory the Duchess was overawed by the size of the place, she must have been excited by the way she became very talkative, she told me their history described how a party of engineers had used cutting equipment to gain access to another factory and it had blown up killing most of them, after that no one tried again, as it was believed the explosion was a deliberate attempt to stop anyone trying to enter the factories and using them.

The two men had automatically gone to the centre of the door expecting each half to open outwards, using the Duchess as interpreter I got them to clear out the tracks and push the doors open from the end, for such a large size they moved quite easily, looking at them now they were fully open I got the impression they were made in position in one piece, they also had flexible sealing flanges all around them, was it to keep air in or gas out. The amount of fine dust stirred up when the doors opened made It difficult to see very far the Duchess was eager to explore but I held her back telling her it was dangerous to breath in such fine dust.

Waiting till we could see halfway down the centre aisle I said its OK to go in, the Duchess didn't know where to look first, darting all over the place like a school kid in a sweet factory she seemed very interested in the machines which I thought was a little odd she didn't appear to be an engineer type, she later told me part of her training had been in technology, but it had been declared obsolete as there was no further need for engineers or research in their society, they had everything they needed.

While she was in such a talkative mood I asked her where the power came from to run the machines, she appeared thoughtful for a minute then her eyes lit up, "if we could get these machines working" she said "we would have enough obligations to last a life time," this outburst shook me it's the first time I've heard any reference to rewards for work, so far the only rewards I've heard about is the one at the food farm when Worra said farm workers who didn't work their quota were sent to the mines.

Walking down the main alleyway we could see the half completed shell of the spaceship hanging on its network of cables, the Duchess made straight for it walking around it she examined it from all sides then she followed the pipe work across the floor to where it entered the trunking, climbing one of the small machines she pointed her Key gun in line with the trunking and fired, this left a burn mark high on the wall which gave us a good guide, making our way over we found a row of small doors which I hadn't seen on my first visit, they were all locked with no visible keyholes or handles but they did have familiar four square patches on each.

Now I wondered if one of the remotes would open any of them, the only trouble was they were hidden around the back of the factory, giving the excuse I wanted something off the car I left them looking for a way break the door down, while I drove around the back of the factory to collect all the remote controllers when I got back they were still busy trying to find a way in, I managed to try all the remote controllers until one produced the familiar clunk, working it a few times to make sure it was working properly gave me the idea how it worked, the power came from the controller which was charged by the sun, a beam of some sort transmitted power to a receiver in the door, and all the time I had been thinking it was powered by batteries hundreds of years old, the Duchess had heard the door open and came running over, she'd have plunged head first into the dark opening if I hadn't caught her arm.

Hold on lets get some lights in there "as if in answer to a prayer the room started to get lighter the lights must be connected to the doors the same as in the spaceship with the lights now at full power we entered the room this had to be the distribution centre, three one inch thick pipes entered from the outside wall, these had smaller diameter pipes branching off them and disappearing through the opposite wall, each pipe had a pale blue light as far as a black obstruction in the pipe, there was no way to turn or move these taps not until I tried the remotes then I saw them physically move and the blue light drain past them, making sure all the pipes were working I looked around there was nothing else to do in here the other doors along the wall all opened into this long room.

The Duchess was looking at metal charts on the wall I went to see why she was so interested in them, she turned to face me, when CLANG, the door slammed shut, by the time my heart had slowed to normal the lights had dimmed and gone out, the glow from the tubes gave just enough light to see where they were, the shock of the door closing had scared me at first, now I felt mad, those two buggers outside had shut it on purpose and now we could hear them piling rubbish against it to make sure we cant get out, I couldn't understand why they were doing this, the Duchess knew why, and soon told me, They think we are sealed up in here and will soon die, then they can claim the credit for opening the factory doors and getting power to the machines, this will wipe out all their obligations and allow them to return to their home planet, and live a life of pleasure.

This was all new to me, I'm getting to learn more about these people every day, so far I thought they were a dull lot with no interest or ambition, now I discover they are driven by the same greed as Earth people. My first instinct was to kick the door and shout for them to open it but that would only be wasted energy, the two chaps hadn't seen how I opened the door they must assume because there were no handles it couldn't be opened, and they hadn't come into the room so they didn't know the other doors opened into this one long room.

The Duchess took in the situation quite calmly and had no objection when I said we have to wait for them to go away, she said "they wont wait long they have to kill Worra as well, she knows you have been inside the factory and was going to show us the way in."

She was sure to run out to the car when it pulled up outside the cabin if only there was some way to warn her, our neck bands could transmit but the Duchess pointed out they wont work in here that's why they lost track of me when I was inside the factory, I couldn't be sure if they would be waiting outside with the stun gun, thinking of some way to distract them I took off my trousers and tunic then threaded my belt through the loops to join them together, the Duchess made no objections when I asked for hers and used them to stuff a passable scarecrow, borrowing her stun gun I gave her the remote to open the door next to the one we used knowing the clang of the door opening would alert them I'd have to be quick, " One, Two Three Go" At the clang I pushed the door open and threw the dummy out as far as I could following it in a genuine Hollywood ground roll prepared to fight to the death.

Only there was no one there, Boy did I feel an idiot standing there in my underpants and vest, but then I chuckled to myself its better to be safe than sorry, running to the front door the Duchess contacted Worra on her translator and warned her about the two men in a car on their way to kill her.

Looking around to find what they had left I found my back pack which had been tipped out and the Tee gun missing, but we had warned Worra and she had my rifle and knew how to use it, in fact I think she'd have no worries about shooting someone, noticing it was getting dark I started to get one of the offices ready for sleeping in, the toilets were still soaking wet from the effects of the extinguishers going off, dragging two piles of mesh from beside the machines I placed them side by side in the office, then I realised we had no lamps and it would soon be dark, now we had switched the power on to the machines perhaps the lights would work.

Collecting the remote controls we tried them against the panels by the front doors, after the third attempt the whole place started to light up and soon it was as light as day, after checking the front and back doors were locked and bolted we went to eat the few biscuit's the Duchess had taken off the trailer and got ready for bed, the Duchess wasn't the slightest embarrassed stripping off in front of me, but I'm afraid I was, and went outside the office to change into my sleeping trousers.

= CHAPTER THIRTEEN =

Pausing to listen to some unusual noises I called the Duchess to ask her opinion, although I didn't need it, as the noise got nearer I could tell it was a pack of yelping dogs in full cry after some unfortunate creature, now it was getting closer we could make out the whirring noise of the car, the two men were heading back hoping to get inside before the dogs caught up with them, or most likely they had seen the lights come on, and were on their way back to finish us off, even thinking this I couldn't let them suffer such a horrible death of being eaten alive by a pack of hungry dogs.

With this in mind I ran to open the doors to let them in, but the Duchess stopped me, she caught hold of the handle pulled it off the spindle and jammed it in the gears, her callous indifference to the men's suffering appalled me, we could hear their desperate screams as they struggled in vain to open the door, the noise of the fighting as the two men fought for their lives reached a peak, I covered my ears to muffle the noise, and it went strangely quiet, but it wasn't the fact I'd blocked out the noise, it was because the fight was over and replaced by the more sinister sound of snuffling and snarling which reminded me of feeding times at the zoo.

"How the hell can you be so wicked" I shouted at her.

"What rights have you to condemn them to a horrible death"

"I have every right "she said in such a matter of fact tone it made me hesitate, She went on.

"The founders gave me the power,. its part of my obligation, anyone threatening my life or authority must answer for it, they tried to kill me by shutting the door on us, its no different to them going in the crusher for fertiliser"

The horrible truth dawned on me.

"You mean to say they are alive when they go in the crusher."

"Of course, They know the risks when they break the law, besides It makes the others take notice."

Now I could see why the other members of the committee had been so frightened of her, a thought flashed into my mind.

"You mean to say when I kidnapped you I was under an automatic sentence of death."

"Certainly, I had no option I had to go with you after the excitement of driving the car and seeing your concern over my safety when we crashed, I decided to go along with you," I was tempted to ask why she fainted when I kissed her but decided not to push my luck, instead I asked if I was still under sentence.

"No she said "under our laws you have redeemed yourself by opening up the factory, it means we can make parts to repair and maintain our own space fleets we wont have to depend on others, as I've already told you our ancestors found many buildings on other planets but they were not advanced enough to exploit the golden opportunity and squandered the technology., the leap forward in space travel without any effort on our part killed all initiative to advance ourselves and we've stagnated ever since."

Her talkative mood had got me interested I was beginning to find out more about their history, so far I've discovered this planet or satellite, is more like a penal colony although there are no guards or warders I think its more like a concentration camp.

Although I suppose that's because there's nowhere to escape to, the inmates seem to work to a self regulated obligation system working the mines and farms in exchange for supplies from the mother world, it could even be a form of national service, all the workers seen to be of similar age, the supply ships that come at regular intervals take back ores from the mines and a few passengers, perhaps it's the time served inmates going home, there must be a command structure that takes orders from a government which is totally in control in a communist like state on the home world, The Duchess said she had authority from the founders to sentence people to death, when I asked her how she had met the Founders she went all evasive, it's the second time I've heard of them, I wonder where they fit in.

Lying in bed fully relaxed thinking about this confusing world it didn't seem real, the whole set up appeared man made or adapted, even the weather was too good to be true, the morning mist provides the water which collects in dew ponds scattered around the countryside and is evaporated by the sun, I haven't been here long enough to see any change in the seasons, by the look of the vegetation they don't have cold weather, although I'm still puzzled about the seals around the factory doors, was there an atmosphere here when they built the factory or did they manufacture it, I suppose by the evidence the builders were far in advance of the people living here now.

By piecing bits of conversation together and guessing, it appears that in the dim past the home world people were helped enslaved or befriended by a superior race who had squandered their resources on space travel and setting up colonies on other worlds, the homeworlders resented this and rebelled taking over the spaceships which gave them space travel but the superior race moved on leaving them far in advance of their evolution.

With all these thoughts going through my mind I must have fallen asleep, because the next thing I remembered was being shaken awake by the Duchess holding out a drink of water for me, this was a surprise, her normal attitude was one of disdain, that's why I call her the Duchess, but this morning she appears in a good mood, maybe she's realised she could go far with me, when she reports to her boss what she has achieved already she would be well rewarded, if that's the case and I play my cards right she will be serving me breakfast in bed before long.

Our meal didn't take long as we only had the food left in my backpack the rest had been on the trailer and knowing the voracious appetite of the dogs there wouldn't be anything left, rolling the door back expecting to see a scene of carnage I was surprised to see no sign of any bodies blood, or ripped cloth, I had to ask the Duchess if last nights fracas really happened or had I been dreaming, the place had been picked clean even the straps on the trailer had been ripped off and eaten by the dogs.

Looking around the battlefield I found the Tee gun which still worked, this meant the two men had soon been overcome, otherwise the gun would have been burnt out, searching around to find if anything else had been left I soon found why there were no bodies, the tracks left in the sparse grass showed where they had been dragged away, no doubt to be eaten by young ones.

Loading the trailer with mesh to make our beds more comfortable with the Duchess driving and me on shotgun, we set off for the cabin with the intention of collecting Worra and returning to the factory, bowling along in the bouncy style of this three wheeled contraption, keeping a sharp lookout or so I thought I didn't see the rope stretched across the path until too late, it caught the front of the car flipping it over backwards, the mesh fell on top of us blocking out the light i felt the mesh being lifted then someone hit me everything went black and I passed out.

Waves of pain washed over me as I regained consciousness the bouncing motion which I thought was the results of my rough handling turned out to be the movement of the car, then I found I couldn't move I was tied up, looking around I saw four men on board and another body lying next to me, fearing the worst I gave it a dig in the ribs with my elbow and felt a glow of relief when it was followed by a groan of pain, at least she was still alive, now the doubts started flooding in, why have we been kidnapped and by who, we were on good terms with the Outlaws and the other people here are too frightened of the Duchess to even touch her.

Unless our captors don't know who she is, this seemed the most likely explanation and there is nothing I can do about it until she tells them who she is, lifting my head I saw buildings in the distance which looked familiar, the food farm my head was hurting like hell and my brain must have been befuddled, it took a few minutes before the combination of being tied up and the proximity of the food farm occurred to me, Bloody Hell they're going to throw us in the crushers, this revelation spurred me into action, bending my feet backwards with the bouncing of the car masking my struggles I managed to undo the rope around my feet without them noticing, untying my hands was a much harder task, no matter how hard I tried by the time we stopped outside the food farm I hadn't made any impression on them.

One of the men jumped off the car and disappeared inside the building, I expect to book us in or collect payment in advance, the other two came to lift me off the car, waiting for one to lean over to catch my feet I lashed out catching him in the stomach forgetting the weaker gravity here I was surprised at the distance he flew before crashing to the ground, the other chap holding my shoulders had hardly moved, I doubled over and kicked him in the face he fell backwards and got tangled up with the driver who had leapt out of his seat to help.

They both sprawled in a heap on the ground, the man I had kicked lay still but the driver ran into the building shouting, the only option for me now was to escape by driving off in the car, this was easier said then done, driving a car with your hands tied behind your back isn't to be recommended, taking a quick look at the Duchess to make sure she wouldn't fall off, I got into the drivers seat easy enough and pressed the pedal, using my chin to turn the handlebars in a wide circle to go back the way we had come.

I nearly knocked myself out when I hit a large piece of stone, this gave me an idea after some distance I turned off the track into a clump of trees going slowly till I spotted a nice jagged rock, I stopped jumped off and started to rub my bound wrists on the rock, engrossed in this I hadn't noticed the Duchess had regained her senses she was banging her heels on the car to attract my attention, when I looked, our captors had shoved a handful of leaves in her mouth and gagged her to keep her quiet.

No wonder she hadn't said anything, of course I had to attend to her straight away by pulling her gag off with my teeth, getting back to cutting my bonds it looks so easy in films, but this rope was made of human hair and it was tough, I don't know how I would have managed if a posse had come after me, perhaps the rough treatment I had handed out damped their enthusiasm, it took ages of hard rubbing before my hands came apart, I tried to undo the ropes on the Duchess but my hands wouldn't work, they had been tied up too long, in the end I had to saw the ropes off with a lump of rock, with the last of the rope off I caught hold of her wrist to rub some feeling back into them she gasped, and I thought she was going to faint again, now was my chance to find out why touching them had so much effect, holding tight I saw her eyes widen as she looked behind me.

"Look the men are looking for us and they still have the stun guns they took off us,", sure enough the four men were coming along the track, they would have to pass near our hiding place, we were well hidden but we couldn't make a run for it, it would mean driving towards them to get back on the track Watching them ambling past one man seemed to be in pain, most likely the one I'd kicked in the stomach, he had dropped behind quite a distance making sure the three were out of sight my plan was to jump out and catch the straggler, to find out why we had been captured, the gods must have been on my side, just as I was preparing to spring he stopped to relieve himself against a tree, with his back towards me. now was my chance to strike. And got ready to pounce.

He was a sitting duck, glancing towards the Duchess I noticed she wasn't bothered, being a gentleman I waited for him to finish, then grabbed him from behind, he squealed but my hand over his mouth muffled the noise, turning him around I motioned him not to make a sound by putting my finger across his mouth.

Using the Duchess as interpreter I asked why they captured us, his answer appalled me he said Worra had told them to kill us, then as the whole story unfolded it all began to make sense, it appeared that when the Duchess warned her the night before, she said two people were coming in a car to kill her, after she got the message a hunting party came by the cabin and Worra asked them to help, they jumped at the chance and had set the rope across the track when they heard the car, our life and death struggle had all been a case of mistaken identity.

This chap was the one I'd kicked and lucky for me he had my Teegun on him, he wasn't feeling too well and was relieved when invited to ride on the car with us, I wanted him as an insurance when we met up with his mates, who we caught up a mile farther on, they were so engrossed in their misery trudging along they never noticed us coming up behind them, it wasn't until our new friend shouted out they realised our presence, of course they panicked and ran to hide in the trees, eventually after more shouting they came out and accepted a lift.

It was nearly dark when we got to the cabin our passengers didn't want to stay the night, so we dropped them off to get back to their village, they didn't seem at all worried about the dogs, this I found quite puzzling remembering my encounter with them and the way they had attacked Valda in the dark, driving right up to the cabin I couldn't see or hear any signs of life, it was getting dark so why wasn't there any lights shining.

= CHAPTER FOURTEEN =

Feeling anxious about Worra I realised she may have gone to the interrogation centre for safety, I don't think she would have gone to the Outlaws village, anyway I was dead tired and hungry, being so hungry must have dulled my sense to danger, no sooner had I opened the door and taken a step inside when I was grabbed by many hands, and wrestled the floor by sheer weight of numbers.

The shock woke me out of my tiredness and in a panic, I lashed out in all directions, feeling my fists crunch into squashy flesh and hard bones gave me a fierce satisfaction, I wasn't going to be carted off to the fertiliser farm again, suddenly the inside of the cabin lit up, someone had used my fag-lighter to light the lamp, now I could see it was Worra she rushed over wrapped her arms around me and said something in her own language, then remembered to switch her translator on.

"We thought you had been killed and when you came in we mistook you for the killers, then she got exited and started talking too quickly it overloaded the translator and it gave out a loud buzzing, the quickest way to silence her was to kiss her on the mouth, this did the trick leaving her with a shocked look on her face, I looked around at my former attackers who had fallen back nursing their bruises, some of them were the first people I met when I first arrived on this planet, now I felt apprehensive who were they after the Duchess or me.

Worra explained, the lights of the factory had been seen from the interrogation centre the people there had gone and found our trailer but no sign of us, so they had come to look for us here, In the meantime we had been captured and taken to the food farm, now I knew what was happening my priority was to get some food, to relieve my hunger our visitors ate their usual handful of biscuits, I wanted a proper meal and started to prepare one, I noticed it had gone very quiet. fearing the worst I hesitated before turning.

I looked around to find our visitors engrossed in what I was doing, it was only a stew but the way they were mesmerised by my actions you'd have thought I was a first class chef by the interest they showed in my cooking, what will they think when they taste it, if their reactions are anything like Worra's it will be a laugh, although at the time I had to stop her eating too much, because of the reaction she had from eating apples at the fruit farm, their stomachs are not used to ordinary food they normally eat only the big biscuits with hardly any taste .

Thinking about the biscuits it seems funny to me why the food farm pays the Outlaws in them, when they deliver the dead dogs, they could just as easily pay them in fresh food, perhaps the biscuits contain drugs or something to keep them mildly sedated, this may account for Worra and the Duchess showing a lot more emotions lately, they have been eating more of my cooking, in fact Worra had been helping to cook which was a lot different to her attitude when I first brought her here, the men are treating the Duchess with respect, that was until she sat down, then they crowded around to watch her eat, then went over to the fire and testily dipped their biscuits in the stew pot to taste it, some of their expressions were quite comical and I could believe it when Worra told me they had only ever eaten biscuits, on the home world they had different colours for different age groups, when she told me this, I felt sure the whole population was being controlled by the ingredients in the food they are eating.

On Earth the British army were supposed to put bromide in the soldiers tea to stop them thinking about women all the time, here they've gone one better and introduced mass birth control,, it also explains why there are no children.

THE BISCUIT BENEFITS

The more I think about the evils the biscuits impose on these people, the more I realise they could be the answer to many of Earth's problems.

We don't make our own individual electricity, so why collect and cook our own food, let factories cook and supply our food in the form of biscuits, we already have a pilot scheme in place, Farley's baby rusks are a well loved large biscuit filled with vitamins to promote growth in healthy babies.

Look at the potential, mass produced biscuits stamped with the ingredients, dates, codes, and information, medicines and drugs could be incorporated for special needs such as diabetes contraception and diets, the nations health would certainly improve, transport and power costs cut, and food waste lowered.

It wouldn't take much to educate the masses to change by using the media to point out the advantages of Biscuits Bringing Benefits Munch when Moving or Dine while Driving, it would certainly appeal to a public who cant afford to waste time preparing meals in their headlong pursuit of pleasure.

Biscuits for famine relief made with high nutrition content and including optional ingredients such as Salk vaccine, Hiv, and other medicines would help feed the masses in refugee camps, made in a modified shape of sycamore seeds they would spin down slowly when dropped from aircraft . and not hurt people on the ground.

The biscuits can be loaded by grain suction equipment for transport and released from aircraft directly over the refugee camps to float down among the refugees where its needed.

This approach will direct food straight to where its wanted. Cut transport cost and stop governments hiving off supplies to feed their own troops, the biscuit aid project would eventually solve the refugee problem, as it would cut the birth rate dramatically.

The Outlaws don't reproduce as they also eat the biscuits, their numbers are kept constant by the introduction of a new member when they swap one who's died, for a living one, and of course the dead body goes into the crusher, the cattle and dogs manage to breed al-right, although they appear to have trouble with cattle breeding, that's why they need our bulls and the rabbits breed, well like rabbits, thinking about more of the puzzle coming together, how these people live, and listening to the snores of the extra sleepers in the cabin I soon dropped off to sleep.

Still feeling dead tired when I awoke I was surprised to find the girls had cleared the men out, and were busy getting breakfast they told me everyone wanted an early start, it turned out the men I had been fighting the night before were engineers from the home-world no wonder they were so keen, to get to the factory, collecting our kit together we piled aboard the two cars and set off, I had already decided to cooperate with them up to a point, I'd show them the factory but I'd keep the space ship for later, I may be able to use it as a bargaining point to get back to Earth.

Our journey was uneventful and we soon arrived at the factory the experts couldn't get off the car quick enough and scrambled to the centre of the door, I hadn't dropped the ratchets the night before, which saved me from showing how to open them.

Going to the far end I put my back to it and pushed, the door slid open the technicians who had been concentrating on the centre crack in the door squealed and bolted into the nearby trees, I suppose seeing the massive door move they thought it was going to fall on them, pushing the door open to its fullest extent soon got them back from hiding in the trees, they were too curious to keep away, they behaved in exactly the same way as I had, when confronted with a brand new factory to explore, running around like headless, chickens not knowing where to look or what to do first, I took charge of them as their leader was still hiding in the trees, I told them using Worra as my interpreter I was going to put the power on to the machines and some of them might start up on their own.

I was still worried about the amount of dust floating about in the air, a spark could set it off especially when I switched the power back on, the best way to get rid of the dust was to open the doors at the other end and let the wind blow it out, leaving the others to their investigations I drove Worra and the Duchess down the length of the factory to the back doors, opening these were no problem in fact it was so easy I let Worra turn the handle, opening them had the desired effect choking clouds of fine dust that had lain undisturbed for hundreds of years came billowing out, and soon dispersed, this made me feel a lot better about putting full power on, while the two girls were looking around in awe at the wrecked ships I recovered two remote controllers from where I'd hidden them.

Now I had the remote controls I could work the lights and get in the spaceship when I wanted to, going back to the power room I started to have doubts, what if these remotes didn't work they must be hundreds of years old, I'd left them in the sun to charge but it was asking a lot for them to still be in good working condition, however they worked first time, I had Worra stand by the door and call out each time I opened a power line no bangs or showers of sparks occurred which was very reassuring.

With all the lights on the machines working, the place looked entirely different, the dark patches had gone, and the whole place had lost its spooky feeling,, The control panels by the machines were lit up and upon examination they looked nothing like I'd ever seen before, there was a packs of cards with diagrams on, which on the machine I was looking at resembled the shape of the finished product lying by the side of it,. the other cards had different shapes which I assumed could be put in the slot in the panel to change the shape of the finished product, this wasn't earth shattering technology as far as I was concerned but the mesh that was used in the process was a different story.

When I had seen it before I thought it was just reinforcing to be coated with plastic or something, having seen what it was used for in the spaceship construction it seemed magical, first of all it expanded as it warmed up, then it served as a frame for the metal to be plated onto, it eventually evaporated leaving hundreds of worm holes which didn't extend through the sides this looked to me as if the shell of the ship was used to contain liquid fuel or coolant, or it could even be sealed in and used as a shock absorber, if the ship was struck by meteors it would dissipate the impact.

To think I had used this marvellous material as a bed to sleep on and for a wick in my lamps its enough to make our engineers on Earth weep at my sacrilege, "Stanlee Stanlee" a barrage of exited shouts broke into my train of thoughts I looked up to see the two girls running towards me calling my name, this must be something unusual to make both of them show such emotion, catching hold of my arms in their excitement they almost dragged me across the factory floor to show me what they had found, at first I couldn't see what they were so wild about, we approached a large cabinet with another box on top, I was puzzled, I had seen this before and there was nothing special about it not until they steered me around the back and I saw for myself what had caused them to lose their normal reserve.

I gazed in amazement at the largest television screen I had ever seen, in my life, this one was at least four feet by three feet but the thickness was only two inches The two girls were so exited to tell me about their find my translator went into overload and just buzzed until I turned it off, the Duchess swung out a seat attached under the panel and sat on it after studying the panel for a few minutes she touched some of the motifs on it, a faint picture glowed into full colour showing a spaceship shell under construction she touched the motifs again and the pictures changed to spaceships in different stages, I was mesmerised and asked if it was to show engineers how to build ships.

"No" they shouted in unison, then Worra continued "this Autobrain can do anything its far in advance of the smaller units we have on our ships, our old records tell us of Master Autobrains like these but we were told they were a myth.

Our experts have said it was impossible for us to make anything so advanced, with our small Autobrains now we can design spare parts and train new engineers to make things and go back to our own self sufficiency, no wonder they had been so excited on finding this machine, if it can do all the things they say it can it sounds as if they are going to save the planet, not believing their claims about it being so marvellous I thought I would play a trick on them, tapping the Duchess on the shoulder to get her attention, I asked if the machine could design something for me.

"Of course it can" she said "just tell it what you want it will understand you, this Autobrain is linking to all the others on the planet and the home-world so there is nothing it cant do," Her matter of fact attitude made my mind up I'd give it something to chew on, I'm not letting a machine beat me, I already had a question I knew it wouldn't be able to answer, I asked it, can you design me a device that will run on its own without using any other power, we call it perpetual motion.

The machine came back straight away with a picture of the Universe, it certainly got the better of me on that one, so I asked the same question again and added man made, this time it showed a ring of eight balls on the end of spokes around a wheel, four balls were missing at the top, I couldn't understand, the purpose of this until the wheel started to spin slowly, the last ball whacked over the gap and sent a shock wave around the circle of balls this whacked the next ball over the gap generating another shock wave each time a ball jumped the gap it unbalanced the wheel making it turn, They say seeing is believing but I couldn't believe what I was seeing, its impossible, of course many inventors have solved the problem on paper, but its another thing to actually get one to work, this Auto brain as Worra calls it is marvellous, its able to think for itself.

The girls were looking through piles of metal plates they had found in a locker under the machine while they were busy I asked the Autobrain to design me another perpetual motion machine, this time it showed ten twelve inch discs on a shaft one eight of an inch apart, half the diameter was immersed in water another similar shaft was interleaved with it, both shafts were supported on the ends, and the discs were covered in fine grooves I couldn't see any sense in this layout, until the machine seemed to sense my thoughts and started it turning, the water changed colour to make it easier to see, then I saw it worked by capillary action, the water crept up the narrow gaps where the discs overlapped, this unbalanced them and they turned in towards each other, amazing to see it working in three dimensional form I still couldn't believe it would actually work, and said so.

"Of course the device will work" Bloody hell I nearly fell off the seat in surprise. I even looked around to locate the voice.

"It will work" the voice spoke again." the machine was talking to me, although I don't know why I should have been so startled, after all I was talking to it, so it was only natural for the machine to answer me, in the same form.

"No" I said "it cant work you can't get something for nothing,

Earthmen have strived for hundreds of years to perfect perpetual motion. In this case air pressure would send water up the same number of surfaces on the opposite side cancelling out," any gain.

= *CHAPTER FIFTEEN* =

The screen went blank for a while, then the machine suddenly fired a question at me that nearly knocked me off my chair again.

"Who are you. What's your Council Member number.

This completely threw me and I was at a loss for words, eventually I introduced myself as Stanley the ambassador to the home World from Great Britain Earth, the machine went quiet then spoke.

"You are worthy Great Britain promoted civilisation and trade to your world by oppression at times, but the oppressed benefited, and showed it, by helping the mother country in times of need, I was impressed by this show of information and wondered how it got so much intelligence, I know spying missions have been visiting Earth through the ages, our history is dotted with details of flying saucers, missing people, and people with extraordinary powers, like Germaine, The Postman and Spring heeled Jack, who popped in and out of our, history, at long time intervals.

I had so many questions I wanted to ask but felt it would be prudent to establish friendly relations with such an intelligent machine first, and the best way is to ask for its name, it had already asked for mine, so I decided to keep our introduction formal.

"Sir you have me at a disadvantage I don't know your name or how to address you." it was the machines turn to be shocked, the screen went blank again, which gave me the impression it was thinking hard, then it lit up and started firing questions at me, "I haven't a name, what's a name for, what does it do, where did you get yours, who gave it to you.

"Slow down I said I cant answer so fast, slow down and I'll answer them", he repeated the questions slower this time, which gave me time to think of some answers, first I said a name is to make you an individual different from everyone else, it also lets people know who you are, and as for not having a name, we cant let you go through life without anyone knowing how to address you, we can soon put that right, thinking hard about an appropriate name for such a clever machine, I ran a few names through my mind, when suddenly the name Merlin came through, putting on a stern voice and placing my hand on the machine console, I started the ceremony of giving the Auto brain a proper name, beckoning the two girls to stand one on each side of me, I started off.

In the power invested in me by the Earth Council, I Stanlee First Ambassador to the Home World Council, formally name this Autobrain, MERLIN, and from this day forth it shall be known by all and sundry by the Honourable name of MERLIN, let no man or woman take it away from you, getting carried away by my impromptu speech I suddenly realised I couldn't put a date to end it off, so I ended with the words," witnessed by Worra Starship Captain and the Duchess of Dar es Salaam.

"There you are Merlin how does it feel to have your very own name, and now you are one of the family, you are named after one of the greatest men who ever lived on Earth, he died many years ago and is still honoured to this day, the screen flickered and a very emotional voice answered.

"I am very honoured to receive such a distinguished name, and to think I am regarded as one of a family gives me great pleasure, I know of Merlin and King Arthur in your folklore, and I consider you to be my King Arthur, looking around to see what the girls were doing I saw they were deeply interested in the proceedings, when I asked the duchess to turn Merlin off she seemed reluctant to do it, and said it would be better to leave him on, they had never seen anything like my conversation before and Merlin would need to gather information from all the other Autobrains he was linked to, I saw the sense in this so after telling him to turn the power off to the screen and only turn it on again when he heard the password Merlin we left him switched on and went to look for the others.

We found them clustered around the half constructed spaceship shell, they were trying to switch on the power to get it working again, we stood watching their efforts for a while then decided to give them a hand but no matter what I did the machine didn't respond to any of my efforts and I must admit I was getting frustrated these people were looking up to me as their leader, their own leader was being ignored after running away when I opened the doors, now I felt I was letting them down.

"Why don't you ask Merlin" a voice broke into my thoughts, I looked at her with a puzzled expression and she repeated herself.

"Why don't you ask Merlin."

"Don't be silly" I snapped at her annoyed at the interruption," how can a machine tell a human being what to do, "Well she replied in her matter of fact voice, they fly our spaceships, of course I'm the silly one I've seen what they can do flying the ship I came here in, apologising for shouting at her, I asked her to give me a hand In putting the questions to the Autobrain.

We made our way back to Merlin, the password brought him to life immediately when we put the problem to him, we soon found out why we couldn't start the machine he told us they had to be synchronized and started up in sequence, previous attempts a long time ago had resulted in very high doses of radiation being released, which had contaminated the whole factory, Merlin said workers had returned three times to get rid of the radiation and the last time they had programmed him to run the factory, but they had evacuated the place in a great hurry and must have forgotten him because they turned off all the power shutting him down, Worra beat me to the next question, when she asked him what are the radiation levels at present and will there be any high radiation levels when the machines are operating and put into production.

I was relieved when he told us they were back to normal the place had been sealed and the radiation had leaked out through the plastic roof over time, that answered the question which had been puzzling me why the factory doors were sealed, next I asked who built the factory and where had they gone, but he couldn't tell me as he'd been switched off and had no idea of time.

Suddenly I knew where they had gone, they were still here with us, in the spaceship outside the back doors, so much for my theories of suffocation rather than die a long painful death from radiation sickness they had decided to commit suicide, Merlin had told us it would take three days for the machines to attain production capability and if we wanted to continue with the spaceship shell we would have to evacuate the factory and run it on automatic because of high radiation emissions from the shell making process, all the other machines were safe to run, we told Merlin we wouldn't use the shell machine as it may damage him, we didn't mention that it would damage us more.

A piercing scream echoed around the factory, and I must admit my first thought was what the hell have they done now, I'm beginning to get as callous as the rest of them here, running to the source of the noise at the spaceship shell we found a party of engineers looking down the pit at something in the bottom, a black column of oily smoke was rising from it which obstructed my view and I couldn't make out what it was, Worra spoke to one of the men who told her a man had gone into the pit to look around and had fallen across a live terminal. and now he was burning.

"Come on" I shouted at the men forgetting they wouldn't understand," help me get him out, no one moved so I so I took off my coat jumped down into the pit and using the weight of the remote control in the pocket flung it around the body, which was visibly burning, I caught the sleeve under it and pulled the body off the metal conductor., Calling out to Worra to give me a hand. I saw two of the men climbing down to help me, this was unusual they normally had to be ordered to do things, between us we got him back onto the floor where we could see he was beyond help.

I felt guilty about his death getting Merlin to start up the machines and switching on the power was my fault, I should have checked first, however I was a little disappointed to find the dead man wasn't their leader, I got Worra to ask where he was, they said he was fast asleep in one of the offices.

This annoyed me, he should have been here looking after his men, I set off to drag him back to work, the others seeing I was in a temper followed me to see what was going to happen, peering around the office door I saw him lying flat out fast asleep on a pile of mesh, on the point of getting a bucket of water to throw over him, I had a better idea motioning the others to keep quiet, I crept in and using my trusty fag-lighter set alight to the mesh it blazed up, he woke up and squealed then the extinguisher went off, I only just managed to get out of the door but he passed me in a cloud of smoke and squishing water, the last we saw of him he was disappearing into the trees, Well, our party just erupted into side splitting laughter, rolling on the floor clutching their stomachs I hadn't seen such hilarity since the village men burst out laughing when Valda attacked me, and I had to fight her off.

Leaving them to get back to their senses I went into the office stood on a toilet bucket and managed to take down the fire extinguisher I wanted the metal dome part for cooking food, while I was loading it in the car Worra came up and asked me why I had tried to save the chap who had fallen down the pit, you don't owe him any obligations and he wasn't one of your weaker sex she said "So why did you risk your own life to help him" there was nothing I could answer to this it was too complicated, I mean on Earth we shoot and injure people we don't know, then do our best to make them better, how can you explain these actions to an alien.

I chickened out and said it was too hard for her to understand and I would tell her more about our way of life another time, The men had followed Worra out and I told her to tell them about the radiation and to take the body and collect more supplies, and if they found their leader collect him as well,, the way they laughed I think he would have to walk all the way, I intended taking Worra and the Duchess into the spaceship outside the back door and didn't want the ex leader nosing around, waiting until they had driven out of sight I drove the car inside and closed the doors.

"Come on", I said to the two girls "I've got something to show you, loading our gear back on the car I drove down the length of the factory the girls kept looking at me expecting me to stop to show them something, Stopping only to open the back doors, I drove out and around the back of the spaceship they both looked at it with puzzled expressions.

"Why are you showing us this its only a piece of wreckage."

Without further ado I took out the remote controller pressed the button and the door opened, they both stood open mouthed in amazement, they had been excited when they found Merlin, now they were speechless, if I hadn't stopped them they would have run forward and plunged inside, telling them to wait while the stench cleared I jammed the door across the opening to divert the slight breeze into the ship, when I switched the lights on there was no holding them back, with squeals they swept past me to stand in awe at the sheer size inside the spaceship I expected them to ask how I managed to get inside. but they were too eager to explore.

= CHAPTER SIXTEEN =

The horrible smell didn't seem to bother them but it bothered me it could be poisonous, at least it couldn't be good for our health.

Grabbing their arms I took them over to the long boxes containing the space suits and fitted them with a suit each they were so exited I threatened to close and lock the door, eventually they calmed down enough for me to show them how to fit and use the air bottles, even then they were taking very little notice of my instructions.

Waiting for an opportune moment I reached across and pinched the little air pipe connecting the air bottle to the face mask, this did the trick it certainly woke them up, snatching off the mask and taking a deep breath of the foul smelling air taught them a lesson they wouldn't forget in a hurry, I pacified them by saying they hadn't clipped the bottles on properly, giving them a remote control each really made their day switching the lights on and off, till in the end I threatened to take them back if they didn't stop playing about.

I told them our priority was to get the air conditioning working as I didn't know how long the lights would last they could fail at any time, Collecting my makeshift lamp and the two girls we entered the lift and I pressed the control panel their sudden .squeals alarmed me for a moment, then, I realised they had never been in a lift before, the sudden rise must have been frightening.

Bypassing deck two I went straight to the control deck, by the tine we got there the two girls had forgiven me for frightening them, They were very interested in the control console and said it was more advanced than the ones they had been used to, Worra used the Auto brain on board to contact Merlin and he told her how to get the life support system working again, she said it appeared to have been turned off deliberately, and she couldn't understand why, if the crew had left the ship for any reason it would have been left on, seeing she was concerned about the missing crew, I told her how I'd found them and took the bodies outside, and I believed the original crew had committed suicide.

She wouldn't believe me she said they could have taken anti radiation drugs to cure themselves, then answered her own question by saying perhaps the high levels they had received made a long and horrible death from radiation poisoning, inevitable, it may also explain why one crew member had fled to the cargo deck perhaps he changed his mind and went to get more air cylinders.

When I told them about the rest of the bodies on level two, they couldn't wait to go and see them,, I tried to put them off, but no matter how gory I made the details telling them they were all bloated up with water running out of them, and if you touched them they fell to pieces, being women the more I tried to warn them off, the more determined they were to see them.

Thinking it might teach them a lesson I gave in and took them down, arriving at level two, now the power was on the cabinets opened at the touch of a button they both grimaced at the first one, and when the Duchess went to pull the face mask off and the body fell forward to land with a squishy sound they both squealed and ran to hide behind me, I didn't laugh it made me feel sick, there's no way I could drag any more bodies in the lift but I still couldn't resist rubbing it in by asking them to help me get rid of them.

Going back to the control centre I said we will have to tell the home world about the ship and they will move the bodies out, The Duchess turned off her translator and spoke to Worra who seemed to agree, then Worra spoke to me.

"We don't need to contact Homeworld, we can use the Outlaws to come and do it for us," this puzzled me, then she explained, if we contact them they will come and take the bodies to the food farm, and get paid for them, this idea appalled me at first, then I could see the logic in it talk about killing two birds with one stone they certainly have conservation working to a fine art here.

“That will solve the problem but how do we contact them Worra came to my rescue again, we have to light a fire and make plenty of smoke its their way of making contact when hunting, after lighting a fire and building a barricade around it with metal sheets to keep out any marauding dogs it seemed a natural thing to sit by it to eat our supper, the peace and quiet had an effect on both the girls and they started to talk about life on their home planets.

The Duchess came from the Home World Worra came from one of the satellite colonies but she had completed her spaceship training on the Home World so she had quite a lot in common to talk about with the Duchess, they were both talking in their own language, but as they had left their translators on I could follow their conversation Worra was talking about her journeys to far colonies, when it occurred to me the vast distances involved couldn’t be covered in her lifetime, the time factor was a dead give-away there was no way she could have travelled so far in the time she said.

“Come on Worra” I jokingly said don’t tell so many fibs, it would have taken you a lifetime to travel there and back, my insinuation really upset her, she jumped up and shouted “Don’t be a fool we don’t travel in time, we travel outside time, when we left Earth you saw the navigation screen, it only showed our position when we took bearings, we don’t actually move we appear at designated coordinates, that’s why Earth people don’t see us most of the time, we are on a higher plane, if we moved on one plane travelling in one of your pathetic chemical rockets we would be wiped out by meteor showers, magnetic storms or radiation bursts.

Seeing how much I had upset her, I apologised and pleaded ignorance by saying “our scientists had told us travel to the stars was impossible without very large rocket engines burning hundreds of tons of fuel every second, just to get out of Earth’s gravity.

And it would take us hundreds of years to cover the distances to the nearest stars“, I said “we were backward by their standards but we are advancing by our own efforts, we didn't get help and training from an advanced civilization, nor did we get space ships given to us", Worra glared at me but the Duchess caught hold of her arm, and shook her head.

"Stanlee is right" she said looking thoughtful for a moment unsure of where to start then making up her mind she began by telling me the history of her people, first she reminded me of the obligations they owed me for showing the way into the factory and the secrets of the spaceship, I suppose it was her way of saying I had been accepted into their society and was telling me about their history as a sort of reward, she told me how in the distant past her ancestors had been brought to the Home World by the Founders from a distant galaxy on a one way ticket, having used the last of their natural resources on the long journey, after establishing the Home World they set about populating other nearby planets which was part of their original plans to seed colonies all over the universe although your Earth was seeded much later.

"What do you mean I blurted out your telling me that Earth was populated by aliens from other planets "Of course she said, your Earth was a recent experiment a lot of different specie were put there, some of the Founders lived there for a while to study you to see if the idea of introducing conflict would increase the advance of some of the specie, the experiment worked beyond belief and civilisation advanced at an alarming rate, The Founders became too friendly with the wine and women which caused the usual conflict, many of the founders who were known as gods were killed along with the tribe of Centaurs who were known throughout the Universe as the most gentle and peaceful of all the creatures, one interesting aspect of the stories in our history is how some humans set themselves up as the sons of Founders.

They had seen how the founders were held in reverence and known as gods so they jumped on the bandwagon as you say, this was a throw back to earlier times when the Founders had introduced religion to recruit the local tribes to construct the navigation beacons, These sons of gods attracted many followers by promising them everything they ever wanted, life in heaven, many wives, journeys among the stars, and everlasting life. Non believers in their religion were dammed to a life in hell, with a fire and brimstone diet, your world is new and has never been given any help by the founders because of the crimes committed against them in the distant past," then she carried on.

"We are obliged to you for showing us the way, in opening up the factory and the spaceship, otherwise I wouldn't be allowed to tell you anything," I was intrigued by her mention of the founders, she had spoken of them before in the context of being in the distant past, now she said she wouldn't be allowed,, as if they are still in control in the present, leaving this question for another time, .I asked her instead "why there were no engines on any of the ships in the scrap yard", she laughed at this, then explained "we don't need engines the whole shell of the ship is a form of engine, fluid is pumped around the porous shell which act as wave guides which multiply the frequencies many millions of times and irradiated by positive nuclei two positives make a negative, this reversal projects the ship and everything inside to the next astral plane, we have to drop back to our own plane, to navigate. by the beacons.

This was all getting too technical for me I could understand tractor and jet engines, but I was out of my depth with wave guides and atoms so I steered the conversation back to their history, they both seemed happier to talk about the past, they said their ancestors who they call the ancients were brought to the Home World by the Founders a civilisation from a distant galaxy who had exhausted their own worlds resources on space travel and were intent on colonising planets on their journey, having settled for hundreds of years and colonising the nearest planets they moved on, the colonists who had resented being exploited rebelled and impounded the space ships taking their raw materials away, this was resolved by the colonists keeping the ships in exchange for raw materials, from the Home World who refined it and sent it to the founders for use in space exploration."

"Earth had been regarded as a stepping stone for colonisation of this part of the galaxy but after the conflict with the founders they cut off all contact with Earth, although we have been monitoring you for thousands of years, the Founders are still undecided if the experiment of conflict is a failure or a success your evolution surges forward in times of war, you squabble and fight at the least provocation, one man can stir up a whole nation to go to war, usually with its nearest neighbour and in the process use up all its resources to ruin each other.

"In the distant past we watched your ebb and flow of evolution with envy at times but comparing it with ours we don't want conflict, no one wins, wars were used to regulate the population, but you cant control your carnal desires and are heading for a population explosion," The Duchess was beginning to sound like Korea Kate the Communist propaganda radio announcer, before she could go any farther I stopped her.

= CHAPTER SEVENTEEN =

"Wait a minute" I said "you have it all wrong, We fight wars to stop bullies enslaving people and stealing their lands, we believe people should have the freedom of choice, which you don't have, here you are told what to do, and if you don't obey, High Council orders without question it's a one way trip to the crushers.

The Duchess was taken aback by my outburst, and fell quiet for quite a long time, I thought I had upset her However I was surprised when she finally spoke.

"You are quite right, perhaps we have misjudged you, While in your company I have been studying you and I must say we have been surprised by your intelligence and manners, our agents gave us a different impression of Earthmen we have found you care for others by the way you tried to save the man in the pit you saved Worra from the bull, and me from the crushers, you even tried to save the two men from the dogs, even though they had tried to kill us, this form of chivalry is alien to us but it is to be commended."

"Next time the high council meet I shall inform them of the details it wont effect the overall plan but it could alter the time scale," this interested me, but she said she didn't know any more, long term programmes were planned by the higher councils, Worra had fallen asleep by now, and hearing the dogs howling in the distance I decided it would be safer to sleep inside the ship.

Waking Worra up, we carried an armful of mesh off the car into the ship, made our beds and settled down for the night, I had to admit it was nicer to have company, the ship still squeaked and groaned but it wasn't so bad, until I remembered our sleeping companions on the second deck, I must have fallen asleep eventually because the next thing I remember was hearing voices, I switched on my translator but still couldn't make out what they were saying, then realised they were coming from outside, by the light from the wedged open door I could see Worra and the Duchess fast asleep, getting up as quietly as possible I crept over to Worra and shook her by the shoulder, she gave a loud squeal and leapt out of bed, either she had been in the middle of a dream or she thought I was about to violate her, the Duchess was awake now and joined in shouting at me, I tried to silence them by putting my fingers to my lips but it didn't work throwing all caution to the wind I I left them, went over to the door kicked it wide open and looked out. expecting to see a group of guards.

Cries of "Stanlee Stanlee" greeted me and I recognised the men outside as Outlaws from the village who had come in answer to our smoke signals, not wishing to start the gruesome task of moving the bloated bodies before having something to eat, I invited them to have breakfast with us, some of them had eaten with us before and couldn't wait to get their friends to try Worra's cooking, it doesn't seem possible they didn't eat the fruit from the food farm but Worra said they had rarely eaten anything other than the biscuits supplied to them I'm beginning to wonder if I'm doing the wrong thing they get all their vitamins from them.

Worra explained what we wanted them to do and they were only too eager to oblige, I showed them where the bodies of the crew members were buries under a pile of scrap and they soon dug them out they were not a pretty sight but that didn't deter them, they stripped the clothes off and shared them out among themselves, these bodies had dried out a bit but I hate to think how they will undress the bloated bodies on level two, leaving Worra to supervise the Outlaws, I had to laugh to myself a few days ago she was terrified of them and hid under the bed quaking with fear, now she was ordering them about as if she had done it all her life, I wonder how she will react when I bring it to her attention.

Collecting my gun and the Duchess we set off to explore the rest of the scrap yard which I found was a lot bigger than expected, it extended far out into the belt of surrounding trees and the vegetation had grown over and covered most of it.

I climbed to the top of a pile and she called to ask if I could see the other factory in the distance, it didn't seem too far away, so I made a note of the direction, having a last look around I got down and headed back to the ship, the Outlaws had loaded all the bodies on to branches and had dragged them off, Worra was looking very pleased with herself she had enjoyed being in charge, the Outlaws had behaved themselves, in fact they were highly delighted at getting a bargain, I asked Worra how their friends will react when they returned to the village dressed in space suits, she said they wouldn't keep them, they would exchange them, for obligations.

This obligation system of theirs seems a good idea it takes the place of money, although ti wouldn't work on Earth you have to be honest, I wanted to go into the factory to have another go on the Auto brain but the girls pointed out it was too dangerous, having nothing better to so we started to explore the rest of the ship, the large crates in the first deck interested me and now we had the power on we could move them to look inside; it was quite simple once we had worked out the sequence, we only needed to get the top box down to unclamp the lid and look inside this disappointed me it was full of granulated rock, but Worra and the Duchess squealed in delight, I couldn't see any reason for their excitement until they explained the rock is only a shield for the metal containers inside, which contain high grade elements of Uranium.

This alarmed me, and I asked them about the dangerous radiation from it, they both laughed at my discomfort, then assured me there was no danger as the metal of the boxes was so dense radiation couldn't penetrate it, this got me interested I told them we use the same system on Earth but its not very effective, Worra looked at the Duchess as if to ask permission then sai

"That's because you don't change the composition before you store it, then change it back to use it, and we don't use lead", she added as an after thought.

The Duchess had counted the number of crates and said it was the same rock being mined at present and very Valuable, as it represented twelve months production, the next two crates held ten sets of space suits in each I called them space suits, but Worra said they were anti radiation suits which made me wonder why hadn't the original crew been wearing them when they had the massive radiation leak, unless they had packed them in the hold and couldn't get at them.

The last three crates in the back row were more interesting they contained boots, tools, clothing and personal belongings the funny thing was it looked like it had all been thrown in together in a panic, sorting it out was going to take a long time, Worra handed me a remote control she had found in one it was different to the ones we had, with a small screen covering one side and the same motifs as Merlin had, Worra said "it may be to access him, through the Auto brain on the control deck," collecting the small pile of remotes I took them outside, climbed to the top of a pile of wreckage, making sure they couldn't be seen from the ground, and lay them out in the sun to charge, I didn't want the party who were due back next day to find them.

Thinking about the remote with the small screen, I remembered The Duchess telling me to leave Merlin switched on as he would be sending and storing information from other Autobrains, if this is the case it means I will be able to contact Merlin from the Auto brain here on the ship, now it was my turn to get exited, I couldn't get back to Worra fast enough, when I asked her if it was possible to contact Merlin from here she said.

"Its never been done before by using t remote controllers."

"Come on" I said "the remotes charged a bit lets go and find out. Taking the lift we went up to the control deck, I was thinking of things to ask, all the question man wanted answers for through the ages, it will be like having the fabled oracles of Rome and Greece at my command, faced with the enormity of power I had, I was at a loss to know what questions to ask, would Merlin know much about Earth history, even if he doesn't I can ask about gravity, the Universe and other worlds, how can we jump from one astral plane to another, reincarnation, hypnotism magnetism, the list is endless.

By the time we arrived at the control desk doubts had settled in my mind, my world wasn't prepared for such a leap forward, I'd rather leave our progress to natural evolution, we are making a right mess of our own world, fighting breaks out at the drop of a hat, half the world is starving, there are too many people and resources are already running out, my enthusiasm had died stone dead, by the time Worra switched Merlin on.

In my fuddled state I forgot to send the password to Merlin to start up and was just about to give up when Worra reminded me and we made contact, I wasn't sure how to phrase the sort of questions I wanted to ask.

In the end I blurted out "Merlin is there a god," I thought it was silly of me to have asked, we are often told by our teachers not to ask any religious leaders, mediums, spiritualists or even swinging pendulums such a question, however Merlin had an answer, "every force has an equal and opposite force, you know there is only one devil, his work is evident all around you, his works are magnified and implemented through the devout and the fanatic, in the past present and future, but there are many gods, every living creature is its own god, this wasn't much of an answer to me so I asked how he came to this conclusion, he started off by saying.

= *CHAPTER EIGHTEEN* =

"In the dark distant past before there was light, all the atoms that make up living things were created, and since then these same atoms have been used and reused many many times, they have evolved and been moulded into every form of life, and until now my present form was a culmination of these trials there is only one you, of you, in the whole universe, everyone is his own god."

At first I thought what a load of old rubbish, then on second thoughts it made sense, our visions of god is man made and for hundreds of years we have been brainwashed into believing what man wanted us to believe, look at the way men have jumped on the bandwagon and called themselves the sons of god, and others flocked to their banners by being promised everlasting and eternal life while being waited on by angels in heaven.

Religious leaders of all faiths are guilty of inciting their followers to kill unbelievers of their faith all through the ages, and calling down the wrath of god and burning in hell for ever on non believers, these religions fighting for peace bring nothing but chaos and conflict even sacrificing their own children and their own lives throwing their gods gift of life back in his face.

Getting these answers off Merlin has aroused my suspicions, the Founders introduced religion to Earth as an inducement to the tribes to build their navigation beacons, so its their fault we have wars and terrible suffering all in the name of religion, the Devil works by duplicity using an adversary's strength to overcome him, that's why only one is needed.

Having doubts about the Founders motives brings a new perspective to the set up here, are they responsible for the drugs in the biscuits to pacify the population into doing as they are ordered, or is the Home World high council responsible,. Worra,s partner waited on the High Council and they ate normal food, the Founders would need them to be normal to govern, so it doesn't prove anything, If Earthmen hadn't killed the Centaurs and Greek gods would Earth be under the control of the Founders now, and eating doped biscuits, now I'm starting to go around in circles and questioning everything, is Merlin controlled by the Founders.

Thinking about Merlin's honesty made me wonder how can you tell if a machine is telling lies, surely I can find out if the Founders are using him by asking him straight out about their future plans for Earth, I couldn't wait to ask him and he answered straight away without any hesitation.

"There are no plans for Earth, it has been bypassed, the natives are too well established and would take up arms, the tribes will annihilate each other, and never reach deep space without help". that sounded a fair answer but I'm still not sure, I could ask him who came first the chicken or the egg, that could blow his gaskets so I won't, instead I asked him "Merlin can you tell me where does our soul come from."

"Yes" he said" its born from your inner self, it was formed over your period of initiation into the world as you know it, its your own conscious perception of thoughts that guide you onto the planes you aspire to,, and enrol you to your next initiation in life, you are your own judge, jury, and executioner, I couldn't help asking how I could be my own executioner, I believed an executioner was a person who ended life, he said.

"It can also mean one who carries out or starts something new", this was beginning to get a bit heavy for me but I couldn't help myself from asking him "what was the purpose of life, what was it all in aid of," he answered straight away just as if he was expecting my question, this brought back my doubts, was I being the victim of a confidence trick, Anyway it gave me something to think about, Merlin replied “the whole purpose of life is to achieve no purpose you came from nothing, and you will return to nothing, no one will mourn when civilization returns to dust, thereby completing a full circle of being and not being, many fall by the wayside and have to start over again on a level to match their sins, The mention of sins interested me it was something I could understand, so I asked him what he meant by sins, he replied sin was anything the intelligent being did against its own conscience or soul, and it pronounced sentence on itself, I wanted to know more but Merlin ended by saying the greatest sin was to reject or take away the gift of life. which was paid for later.

Talking about the gift of life gave me an idea, and I soon put the question to him that would soon show if I was being fooled by the founders or whoever ran this place.

“Where did the people come from who made you” he surprised me by answering right away, I thought he would have made an excuse to evade the question.

“They the Founders came from a galaxy outside your present solar system and travelled through time, but through inter breeding with other races and evolution they lost that ability, just like this colony has lost its ability to evolve, if you hadn’t appeared this planet would have failed,“ This statement shook me, it meant there was a governing ruling body in charge, somewhere.

It even hinted the whole universe was governed by an entity of some sort, Merlin had said there wasn’t a god, unless the multitudes of gods he mentioned linked together like atoms, and made decisions to bring order to the universe, I was beginning to think I was out of my depth, I was getting in too deep, each question I got answered only led to more questions.

Looking around for inspiration to change the subject, I noticed Worra wasn’t taking the slightest notice what I was doing, this seemed odd compared to the excitement she had shown when we had first found the main Autobrain in the factory, I asked her why she wasn’t interested, its all old history to me she said, I thought she looked so bored I’d wake her up, I don’t know what put the idea in my head, but I asked if she would like to speak to some of her old friends at the space training school, she looked at me in amazement then shook her head that would be nice but its impossible these machines can only plot courses and design things how do you know I said if you don't try, she still didn't believe me so I said just ask Merlin to send and receive pictures from this locality to your old training station and see what happens.

She did this and I shall never forget the look on her face when a picture of a workshop appeared on the screen, she squealed when a startled face appeared only to disappear, then came sounds of shouting Worra stood with her mouth open gazing at the screen, I was puzzled at what was going on then realised they had never seen pictures before only navigation charts, and for trainees to be suddenly confronted by pictures must have been traumatic, I could imagine them telling the instructors to look for themselves, and dead on cue three faces appeared, staring back at us.

Worra gave another squeal she had recognised them, and started talking so fast it overloaded my translator, the way they were chattering away I knew I wouldn't get a look in so I returned to the loading bay where the Duchess was still sorting out the contents of the crates, I sat beside her and told her we had contacted the Home planet and got pictures back she didn't believe me at first but when I explained further she got excited and told me.

"In their ancient history there were stories about pictures on the Major Autobrains but they had been dismissed as another ancient myth, now we had switched on Merlin he had linked all the units together and gave them his capabilities, "we can contact other colonies, and be more self supporting," no one had mentioned this before as far as I can see they are already self sufficient, and running a successful trade.

"No the Duchess explained we are dependent on the Home world for all our training and repairs soon we will want them to recharge our generator it is nearing the end of its life, now we have the supply of ore from the spaceship we can fulfil our obligations," I was beginning to understand more about this system of Obligations its more like bartering.

I can't understand how they manage to keep any records of

transactions unless its all beamed to a central records office, if these Autobrains can plot courses through millions of miles of space where worlds move in different orbits all the time they certainly have the ability to do accounts, that reminded me now Worra is using the Autobrains as a telephone service, the accounts people wont be too happy, I asked the Duchess about this she told me there were some units stored in the factory that had blown up the explosion had ripped the doors off, but the units were packed in crates, and as far as she knew they were still there, I asked her why she hadn't mentioned this before if these marvellous machines were mentioned in your history why hadn't you used them.

"We had no use for them" was all she said," and its forbidden to go there," a little thing like that wasn't going to stop me and I asked her to organise an expedition, she did this and told me they would be here in two days, bringing more supplies and men, some were to stay here and survey this place and the rest to go with us to the other factory, this meant they would be a day late coming to us, we had plenty of supplies so it was no problem.

While eating our midday meal I asked Worra and the Duchess what else would I find in the wrecked factory, they said they didn't know but Merlin would, I still can't get used to the idea of asking a machine questions, but this proved a winner, Merlin said I would find seventeen complete Autobrains, inside and three outside, still packed in their boxes.

= CHAPTER NINETEEN =

Then he carried on saying there are seven fliers outside the back doors, these had been made in the factory and there were others in stages of construction Fliers was a funny name to call something and I asked him what a flier was Merlin went into a long description of a machine that flew through the air I thought he was talking about spaceships until I asked him to show a picture of one.

I should be used to shocks by now but this took some beating, up came a picture of a one man helicopter, it looked incredible and cheaply made I wouldn't expect it to fly, two A frames joined at the back with a rudder, and a ring joining the ends of the rotor blades, with no tail rotor it looked like the rotors were driven by jet engines, my initial excitement was soon dampened when I saw these pictures the thing were too primitive to be any good, feeling disappointed I turned Merlin over to Worra knowing she would be a long time on it, especially if the Duchess wanted to talk as well, I decided to make a list of all the questions I wanted to ask him in future, looking through my back pack for a sheet of paper I found my copy of the Beale Treasure Code which I had been trying to decode for many years.

Actually the whole story is that a group of adventurers found gold in West Virginia USA in 1820, for security they buried it and left a box and instructions with a local hotel owner that if they didn't return within ten years to open the box.

When no one turned up to claim the box, it was broken open, it contained a letter and details about a seventeen million dollar treasure, buried in West Virginia United States, instructions which were in three sheets of numbered codes described where it was buried each number represented a letter, page one has been decoded by numbering the words of The Declaration of Independence and using the first letter of each word the message revealed the contents of the hoard buried in Bedford County VA, Page Two, giving the exact location of the seventeen million dollar treasure has so far defied all attempts to decode it, page three gives the names of next of kin who share the money, I've been working on these codes for years so far without success, the trouble is parts make sense and this encourages you to spend more time on it, I believe the American army use it for training purposes, if they cant find the answer I don't think I stand much chance, anyway I console myself with the thought if I found the solution there's no way I could go to the States and dig it up, I'd need seventeen million to pay for security.

Taking a quick look at my latest attempt to solve the codes gave me an idea, if Merlin can work out numbers and distances and can think for himself he should be able to solve a simple thing like a code, gathering the papers together I told the Duchess to carry on then headed for the lift, arriving at the control centre I found Worra gazing at a blank screen, in my excitement to get Merlin working on my code I shouted at her for shutting it down, she looked on the point of tears, asking her what happened she said the training centre had become frightened and switched off.

This suited me as I'd been wondering how to get her off the line, it's the hardest thing in the world to get a woman to stop in the middle of a conversation, and its sure to be the same here, Of course I said it was a shame, perhaps in a day or two they may get used to the idea of contact especially when the salvage parties get here and they want to contact base, she brightened up at this and when I said she would be in charge of communications it made her day, reminding her not to tell anyone the password for switching Merlin on, I explained that this way she would be in control, If her bosses wanted an explanation she could say she brought Merlin to life and he thinks she is his mother, she laughed at this and seemed very pleased at me taking her into my confidence.

Now I was at a complete loss I hadn't given any thought to the way I could present my codes to Merlin and had visions of taking hours to dictate lists of numbers into his memory, Worra noticed my predicament and asked me what I wanted to do, I explained all about the codes and gave her paper number two, this was the one giving the location of the treasure, she gave Merlin a string of instructions then pressed the sheet of paper to the screen, this didn't make sense, for a start the numbers would be the wrong way round and how could a machine read, doubts began to creep in, I think I've been carried away and I'm expecting too much, tapping Worra on the shoulder I said.

"Leave it till tomorrow it will to give him time to sort it out.". "Look," She said, there's something coming through now", she couldn't read it as it was in English, I was amazed it was only one line, but I had been working on this code for years and now this square heap of tin with a television for a head, had worked out the first line in less than a minute, we waited another five minutes but nothing else came Worra said the rest of the code must be harder to decode and will take longer, I looked at the words on the screen mesmerised, they read OUR PIT TRYST. that would certainly make sense to me, although many times in the past I thought I had the answer it pays to check, so I asked Worra to get an explanation how Merlin got this message, a whirring noise and clonk answered me, a small sheet of metal ejected from the base of the screen, I picked it up turned it over and printed on it. was the message.

OUR PIT TRYST, complete with the key to decode it, I cant believe its so simple just add the code number digits plus seven and swap the results to the alphabet, now I'm puzzled why hasn't Merlin decoded all of it, and why are there two keys, I get the impression the code is a game with many keys hidden in the decoded bits to give a clue to other keys, these two keys use addition, perhaps others use subtraction, multiplication or division.

THE BEALE TREASURE LOCATION CODE

CODE KEY No ONE	CODE KEY No TWO
71 = 8 + 7 = 15 = O	
194 = 14 + 7 = 21 = U	
38 = 11 + 7 = 18 = R	
1701 = 9 + 7 = 16 = P	
89 = 17 + 7 = 24 = X	89 + 17 +7 = 113 = I
76 = 13 + 7 = 21 = T	
11 = 2 + 7 = 9 = I	11 + 2 + 7 = 20 = T
83 = 11 + 7 = 18 = R	
1629 = 18 + 7 = 25 = Y	
48 = 12 + 7 = 19 = S	
94 = 13 + 7 = 20 = T	

I kept looking at the results to make I'm not seeing something that wasn't there, surely someone would have seen this over the last one hundred and fifty years, many cipher experts agree the codes have no intelligent messages in them, but I believe even this first bit may start them thinking.

The story says there were thirty men who agreed to bury their gold in a pit in West Virginia, Merlin mentioned a pit. so I reckon he's on the right track. If I remember correctly it also said a letter would come from St Louise in ten years time. Containing the keys to unlock the codes and it uses the letter S which is number 7 or 19

At that unruly time 1832. there wasn't a cat in hells chance of mail deliveries being guaranteed, it appears that part of the story is a deliberate diversion, perhaps the letter referred to may have been the letters in ST LOUIS changed to numbers and used in adding to the code numbers, in fact I don't think there ever was any letter to be sent from St Louis.

It would certainly be ironic if the letter they had waited ten years for, had been in their hands all the time.

It was getting late so I decided to have my supper and then turn in we were sleeping in the cargo hold, I suppose it was because we had left our beds in there and had no reason to alter the arrangements, the nights were very pleasant and the sight of myriads of stars through the open doorway was awe inspiring, although I couldn't recognise any of the constellations, it didn't distract me from admiring the heavenly display, I thought what if the worlds on this plane occupy the same space as the worlds on another plane, would that account for ghosts poltergeists' and lots of other things, I mean there are lots of stations on a radio and they all originate and are carried by the same means, until they drift and overlap they wouldn't know each other existed.

The two girls couldn't understand my curiosity of gazing at the stars, they showed no interest in them at all, when I asked they said their ancestors had visited some of them and most were just barren worlds, even that got me interested, they had said most of them, to me that means some of them wasn't barren, I had lots of questions buzzing around in my head crying out for answers and I'm sure the girls knew this, when I started asking about anything they turned off their translators and pretended to be fast asleep.

The sun streaming through the open door woke me up quite early, the other two were still asleep, this suited me as I couldn't wait to get up to the control deck to find out if Merlin had deciphered any more of my code, the Autobrain was still humming away when I got there and there was some writing on the screen, I wasn't sure how to get a written copy so I wrote it down on a piece of paper, this time it read HID A TROVE, and showed how Merlin had worked it out, this must be another key to the codes but its nothing like the other two, which points to my idea that there are many keys and each only decodes a small part..

Code Key No Three is used here and later.

83= 11
1629 = 18 + 11 = 29 + 5 = 34 = H
48 = 12 + 18 = 30 + 5 = 35 = I
94 = 13 + 12 = 25 + 5 = 30 = D
63 = 9 + 13 = 22 + 5 = 27 = A
132 = 6 + 9 = 15 + 5 = 20 = T
16 = 7 + 6 = 13 + 5 = 18 =R
111 = 3 + 7 = 10 + 5 = 15= O
95 = 1 + 3 = 17 + 5 = 22 V
84 =12 + 14 = 26 + 5= 31= E

HID A TROVE, this looks more like a breakthrough although I still haven't any idea how Merlin does it, how can a machine work out so many combinations of numbers, there must be billions of different ways and he did it in such a short time, if I ever get back to Earth I'd like to take one of these machines with me, having finished copying the writing on the screen I tried to decode more Numbers.

But they don't add up the same as the first lot, 48 = 12 + 18 where does the 18 come from Then I saw it the preceding code number added up to eighteen so you just added that in, I can't understand where the five comes from, it could be the missing letter U from St Louis, which is 21 or 5, and used to break up any pattern in the codes to make it harder to decode, I suddenly felt dismayed looking at the results I had so far, I discovered the numbers 1629, 48, and 94, overlapped, They were used in both solutions, this couldn't be correct, never mind I've had so many disappointments working on the codes, one more won't make any difference, besides I've got something, to work on which I never had before and Merlin will provide me with more answers.

Collecting my notes I went back to the cargo bay where Worra and the Duchess were having their breakfast, both were glad to see me, they thought I had gone outside and were getting anxious when they couldn't find me, I said I was sorry for not telling them, but I was too interested in what Merlin was doing for me and it took longer than I expected, Worra reminded me the work party would be returning soon, and unless we moved back to the factory, they would find we had access to the spaceship.

I'd forgotten about them, being preoccupied with using Merlin, eating my breakfast in quick time, we left the ship and moved our stuff back to the factory, this started me thinking, the newcomers would want to use our new toy for communication I told Worra to switch the screen off until they got used to the idea she was the only one to use it and not give her password to anyone.

Two car loads of engineers arrived later that day, I couldn't count them, as they jumped off the cars and swarmed into the factory like exited schoolboys, I wasn't expecting them to be so enthusiastic, it was only later that I found they were from the home planet and were keen to be the first to find out something new, to earn themselves obligations, I don't suppose there was any chance of getting any on the home world.

The fellow in charge of them wore a translator so I could talk to him without any problem, He appeared more intelligent than the others but not very advanced in technical science in fact most of their training seems to be very poor standard, the home world teachers must have orders not to teach them too much, Rudo that was his name lived on one of the satellite colonies and had been picked to train on the Home World, although they didn't study or have any contact with Home World students which confirmed my theory there is a two tier system, in that case I would have sent the higher trained, team, perhaps those in charge believed this lot would blow themselves up, especially as they were dealing with a factory similar to the one destroyed years ago which killed their best engineers.

I suppose it makes sense to send in the second team first, it wouldn't matter so much if they were lost, Rudo their leader said he had brought twenty men with him, I thought with me and him that's just enough to make two football teams, I laughed at this idea at first, although looking them over they didn't seem a very tough looking crew, in fact all the people here are a skinny lot, never mind a few games of football could work wonders, I remember how the British workers building the railways in South America taught the locals to play football, now they have some of the beat teams on Earth.

Once I'd started thinking about football I couldn't get rid of the idea of teaching them to play, it might give them a bit of aggressive spirit, shake them out of their lethargic way of life, something to live for, there's plenty of flat ground around about, and we only need a ball, we used to use a pigs bladder blown up inside a leather panelled ball, I can ask the girls to sew the panels together then we have to make a bladder out of something air tight I suppose a dogs bladder can't be much different.

The girls, Christ I'd forgotten about them, Twenty Two men and only two girls that's asking for trouble, they hadn't been any trouble so far, only these newcomers are from their own world and I don't know how they behave between the sexes there,. If the worst comes to the worst I can always get one of the girls to fall for Rudo that will defuse any trouble, The men had started to unload so I took him for a tour of the factory, he was interested in anything and everything, I showed him the suspended shell of the spaceship which fascinated him, I asked if they had brought any dark glasses with them as the process of working the machine produced blinding white light, he said they hadn't but it would be easy enough to use smoked glass or powdered carbon between sheets of glass. as soon as he said this I knew we would get on well together, it showed he could use his initiative.

While showing him around the scrap yard we bumped into the Duchess, I introduced them and Rudo bowed to her, this was the first time I had seen anyone do this and I was at a loss to know what to do for a while, although it seemed natural for them, Later I asked Worra if this was the normal thing to do like we shake hands on Earth.

"Certainly " she said "The Duchess is a member of the high order and we have to respect this on first meeting her, failure to do it, or to harm her would result in a termination," I should have remembered this from the time she told me I was under sentence for kissing her when the car overturned,. I don't think I have any worries about the men messing with her if the threat of the crushers is the end result, Worra could be different going by past encounters I think I am worrying over nothing, the men don't show the slightest interest in women.

My next problem is how to make a football, all I can think of is to shoot one of the dogs that come around now and again, then use its skin to make the panels and its bladder to inflate it, Thinking about killing an animal just to make a ball doesn't seem right the best thing I can do is send a car to the food farm in a couple of days to collect some bladders and skins and any more supplies we may want,. thinking of the problems coming to light almost made me give up on the idea.

Making a football is a lot harder than I thought after I've cut the skin to size and shape I have to make the needles to sew it together. get hair for cotton then make a pump to inflate it, Worra and the Duchess don't seem the normal domestic type, I can't imagine them sitting down sewing. The clothes here all have welded seams, the material feels like plastic, but when you hold it up, the light shines through millions of little holes in it, That's the reason I cant use it to make a ball, its not airtight, Merlin that's the answer, he was made to solve problems so lets give him some, he can work out the size and shape of panels, he may even have a list of football rules, the more I think of the game the more I realise I don't know much about it, The offside rules for a start are all double Dutch to me.

Walking back through the factory I noticed the newcomers were busy sorting out how the machines worked, the Duchess was working well with Rudo, she was relaying instructions to Merlin and they had quite a pile of metal plates by them, this reminded me there could be some for me when I asked she said "there's two waiting on the other machine" I knew what she meant, we had left the auto brain on the spaceship switched on, This information made me forget about football, I was so eager to collect my messages I forgot to pick up the remote control, from the office where we had set up lodgings and had to go all the way back for it.

The door of the ship was on the far side away from the factory door this meant if anyone strolled outside they wouldn't see me, Closing the door behind me just in case, I soon noticed that horrible smell of rotting bodies, in future the air pumps will be left running, I also realised the silence was threatening to close in on me, it felt so sinister I was scared, if there hadn't been a message for me to collect I would have gone back to the factory, however I carried on, collected the message, and left the ship.

Once outside in the fresh air I felt a lot better, even laughing at myself for being scared and putting it down to the fact that before I'd been too busy to notice, and the last few times I had company, being so occupied with my thoughts I didn't read the message until I got back to the office, after reading it and seeing partial solutions I get the impression the codes are a game, most experts say there are no message in them, but when you decode one part it gives clues how to solve more, I can't see how a human mind can ever solve them, even Merlin seems to have been beaten, anyway I'll give him more time although if he does manage to find the answer and I ever get back to Earth there's no way I could go and dig up a grave, I understand treasure hunters who think they have the code solutions have been jailed for digging in cemetery's in West Virginia with mechanical diggers.

Looking at the latest solution I saw a different key was used this one included subtraction, this meant my theories could be right.

63 + 9 = 72 -- 7 = 65 = M, KEY 3
132 + 6 =138 - 7= 131 = A... .KEY3
16 + 7 = 23 --7 = 16 = P KEY3
111 + 3 =114 -16 = 98 = T, KEY3
95 +14=109 -16 = 93 = O, KEY3
84 +12= 96 -16 = 80 = B, KEY3
341 + 8=349 -16 =333 =U, KEY3
975 +21=996-16 =980 =R KEY3
14 + 5 = 19 +16 = 35 = I, KEY4
40 + 4 = 44 +35 = 79 = A, KEY4
64 +10= 74 +16 = 90 = L, KEY4
27 + 9= 36 +90 =126= V, KEY4

This is getting interesting each bit decoded gives a clue to another key to be used later, so far the letters are encoded by adding or subtracting from the line above it, and decoded by reversing the plus and minus signs, if I remember right Edgar Allan Poe used a similar idea with six keys to encode one of his famous ciphers, if I return to Earth I may be able to prove for certain he wrote the Thomas Beale hidden treasure codes.

= CHAPTER TWENTY =

Busy sorting out the intricacies of the codes I completely forgot about the expedition to the other factory and it wasn't until one of the men called me that I realised they were all ready to go, collecting my back pack, gun, and anything else I thought would be useful I jumped into the driver seat and we set off.

Leaving Worra behind still worried me a little, the Duchess tried her best to reassure me saying it was unknown for the sexes to harm each other here or on the Home World, although she knew it was different on Earth, Worra had told her about our volatile attitudes between the sexes, and the love hate relationships we go through once we are married.

Once she had broached the subject she became more talkative on the journey and told me why sex never bothered them, In the distant past the founders decided sex and drugs were destroying their civilised society, they had to act fast before it was too late, first they isolated all the drug takers onto mining satellites and put them to work in exchange for food and drugs, gradually reducing the amount in relation to the work.

This resulted in a workforce too busy working hard to get more drugs, they didn't have any time to cause trouble, finding the idea worked beyond their wildest dreams, the Founders extended the experiment into birth control to cut the birth rate, by making the population dependant on a food supply of biscuits laced with different drugs vitamins and anti radiation doses.

This had been introduced over many years and so gradual the public didn't notice and were content to live in a happy state, this idyllic lifestyle was perfect for a while, then it was realised the population had lost desire to better themselves, there was no need to learn a trade, no one bothered to train as engineers or teachers, they had everything they wanted even space travel, this life without competition stifled the advancement of evolution and the whole planet went into stagnation, the Founders are not allowed by their own laws to interfere, but historians believe they increased the potency of the drugs on the home world which killed off or terrorised the drug addicts into giving up.

This solved most of the problems but now there were no teachers to teach, some were brought from Earth a long time ago but proved mentally unstable through religion and were returned, over time trade between the Home world and satellite colonies became established resulting in a drug free Home World, training people from the colonies in exchange for raw materials, this is the system we use now and its working very well.

I believe this system would work very well on Earth,, put the druggies in a concentration camp and use confiscated drugs to pay them to stay there, I'm sure it would work out cheaper, We could also use the do gooders who are sure to complain about such drastic measures as camp guards, if they are so interested in the inmates welfare they would be the ideal people to look after them, Just imagine, it would cut the homeless figures, make crime almost non existent, empty hospitals, and best of all it would neutralise the do-gooders.

A shout brought me out of my day dreaming. they had sighted the factory in the distance and of course everyone craned their necks to have a look, we could see right through to the other end, as we got closer we could see the massive door lying on the floor, we couldn't drive right in the door was blocking the way, Pulling up alongside it I wondered if we all got together we could move it, lining up the men to lift the door I noticed large holes had been scooped out underneath, it wasn't until we had lifted and pushed the trailer under it, did I realise what had caused the holes, when the explosion had blown the door off some people had been crushed underneath, and they had been left there. The wild dogs had smelt the blood and had burrowed in to get at them, We found bits of clothing and a few metal possessions but no bones, pushing the other trailer under the other side we were able to move the doors right out of our way.

I noticed none of the chaps appeared very eager to explore the factory, it may have been the smell of the dog mess lying about but it didn't bother me. I was interested to see what lay behind it, so I drove straight down the central alleyway, passing the machinery I saw it was different to the first factory these were smaller and there were a lot more of them, We couldn't drive straight out as the door was hanging over at an angle, in fact it looked very dangerous, we couldn't leave it like that if we were going to work anywhere near it, I couldn't wait and crawled under it to see what lay on the other side, and what a sight it turned out to be, this was a proper scrap yard, piled high with broken machines of all sorts as far as we could see, I recognised some cars similar to the ones we came here on, they didn't look damaged in any way.

Rudo had followed me out, and gazed in awe at the amount of material lying about, look at all this" he shouted in excitement, . "we never knew this much existed on the whole planet, we can repair it and be independent of the Home World," This was the second time I had heard this and it disturbed me, I didn't want to be part of any revolution, He watched me poking around in the piles of discarded machinery and asked if I was looking for anything in particular, so I told him about the flying machines the Duchess had described, I know all about them he said, they were part of our training on the Home World.

"Great" I said this was a stroke of luck "you can tell me all about them and how they work," Oh no" he said "that's the trouble, they don't work, as part of our training we had to explain why a heavier than air machine cant fly and prove that stories of flying machines are a myth" I was amazed these people living with space travel almost every day are being led to believe they can't build a machine that would fly, do I tell him about our aeroplanes on Earth, or wait and see if I manage to get something flying, it will make me a god in their eyes, However Rudo soon brought me back down by asking me to turn the power on to the factory, it would soon be dark and by the look of it the wild dogs had been roaming freely through the factory at night.

"We can use the side rooms to sleep in" he said, "but he didn't fancy waking in the morning and running into a pack of wild dogs," This factory had been built on the same plans as the other one which meant we had no trouble finding the power house, before switching the power on, I warned Rudo about the accident at the other factory when the chap was killed after falling into the pit and electrocuted.

We both checked around the machines and warned the men to be careful then returned to the power house only to find the doors wouldn't open when I used the remote controller, i felt a proper idiot it had worked al-right before, it was no good trying to break them open the metal was too strong, I wondered if the place being open to the elements had corroded it shut, we could hear the bolts retracting back and forth but it still wouldn't budge, all the doors in this place were a tight fit, the slightest corrosion would stop them opening, we could do with a flame cutter or heat of some sort if we heat the door when it cools down it will free itself, In desperation that's what I decided to do, so sending the men to collect wood and anything that would burn we piled it against the door and set fire to it.

We let it burn for a while then when it was too hot to touch we threw buckets of water over it working the remote all the time, the door suddenly clicked and swung open taking us all by surprise, clearing away the remains of the fire we entered the room.

There appeared to be a lot more electrical conduits in here, which accounted for the higher number of machines, Using the remote to switch on the main power, I showed Rudo how to switch the others on while calling out to the man outside as each conduit gave out the familiar glow the main lights came on first time which I thought was marvellous they had been open to the wind and weather for hundreds of years and yet they came on straight away, or rather they glowed dim at first then gradually became brighter I expect they had to warm up, on switching them off they still glowed for a long time, then gradually dimmed and went out.

Leaving Rudo to carry on I went outside to continue my search of the scrap yard only to find it was getting too dark, so after collecting my bedding off the trailer I prepared my bed for the night, The Duchess chose to sleep in the same office and the men spread out into the next three rooms, we had to cook our meal in the factory otherwise it would have set the fire alarms off, finishing our supper we all retired for the night only to be rudely awakened by a horrible snuffling noise outside our door, in my befuddled sleepy state I couldn't make out what it was at first then I recognised it, the dogs had been used to roaming through the factory and had been attracted by the small of cooking, not wanting that noise keeping me awake id have to put a stop to it.

I had an idea, putting my trusty fag lighter on maximum to scare them off, I edged the door open, CRASH, the weight of dogs hitting the door hurled me on my back, suddenly they were all over me and I thought my time had come, luckily I'd pressed my lighter and the flash of flame distracted them from attacking me.

Next came a bang and Whoosh a torrent of water flooded the place, Well the dogs just disappeared, how they all got back out through the door I'll never know, I just stood there dripping wet looking at the Duchess who stood speechless I don't think she realised what had just happened the shocked look on her face disappeared when she saw I was laughing, the sight of her right royal majesty; standing bedraggled like a drowned rat, water streaming down her face, made me burst out laughing.

I couldn't help myself it must have been the relief at surviving such a perilous situation, thinking about it now I must have been dead lucky, there must have been five or six dogs in the room, we wouldn't have stood a chance and even when the extinguisher went off if one of them had knocked the door shut, they would have been locked in here with us, with no weapons to fight them,,,,. we wouldn't have lasted long,. In future I'll make sure I carry a Tee gun at all times.

Making sure the dogs had gone I went outside to collect some of the wood we hadn't used to heat the power house door, I could hear the barking of the dogs receding in the distance, they were having their usual running fight, The dogs fleeing from the explosion in the office had collided with others trying to get in which give them an excuse for a royal battle.

A nice blazing fire in the office soon warmed the place up, none of the other chaps had come out to see what had happened which seemed a bit odd, then I remembered looking after number one is the normal thing on this planet, besides it may have been all for the better the Duchess had stripped off and after washing her clothes in the toilet sink was now busy hanging them out to dry, out of modesty I kept my pants on then thought better of it washed them and hung them up as well.

While waiting for Our clothes to dry the Duchess came and sat beside me, I couldn't help but look at her she had a nice trim figure and looked about twenty, but I knew she must be a lot older, I was tempted to ask her age then changed my mind we had a good relationship and I didn't want to spoil it.

After a while the fire died down the room warmed up and our clothes had dried, so I cleared the embers outside closed the door and resumed my broken sleep, Our nocturnal adventures must have tired us out and I didn't awake till late, and then it was the weight of something pressing on my arm that woke me, I was shocked to find the Duchess cuddled up to me, she had evidently got cold in the night and decided to have a warm, when I moved to get up she wasn't the slightest embarrassed on finding herself in bed with me, I made my excuses and left, as the proverbial News of the World reporter would have said, it still shocks me the casual attitude the people here have regarding nudity.

Quickly dressing and eating a hasty breakfast we emerged from the office to find the workers busy evaluating the machines. And getting ready to run them, although I was eager to get outside and explore the scrap yard I couldn't resist stopping to see what they were doing.

The method for turning out the different shapes of mesh soon showed it was quite simple, the Duchess had used the Autobrain to design the programme on a postcard sized plate that emerged from a slot below the screen, this was taken and put into a similar slot on the appropriate machine, the operator pressed the motifs and after a few minutes a shaped and sized mesh panel dropped out onto a small trailer, I expect this would be taken to the plating process where it would be finished off.

The Duchess had been following me and she naturally accompanied me outside I thought she would be handy to recognise any flying machine parts amongst all the wreckage.

We soon found three cars which looked as if there was nothing wrong with them, I know they had been lying out here for perhaps hundreds of years but I still tried the controls just in case, none of them worked which reminded me I've been using these machines for some time now, and haven't a clue how they work, now was a good chance to find out, clearing some sheets of metal and other debris off them, I managed to drag one out into a clearing, tipping the car on its side I started to check it over, engrossed in finding out what was wrong I was startled when the Duchess called me, to look at some more cars she had found, these had all been covered by scrap metal and looked like they had been covered on purpose, this gave me another puzzle were they hidden by the original builders, or by the chaps killed in the explosion, digging these out would be a hard job so I left them for the team to sort out.

Advancing deeper into the piles of scrap made me wonder where it all came from, it all looked the same age and similar construction, so it hadn't come from other alien planets which disappointed me, there were no tracks of any sort although they could have been grown over by now, the only clue I could find was the way the scrap overlapped, it all seemed to have been dropped from above, This could confirm what the Duchess told me about flying machines, but why did she have a living memory of them when Rudo was taught such things never existed.

I could hear the Duchess throwing bits of tin and piping out of her way as she burrowed into a large pile, Stanlee Stanlee Her muffled shouts sent me running in her direction, she only used my name when she was in trouble or excited.

"I've found one "she shouted as she heard me clamouring in behind her, catching up with her, I examined the object causing her excitement, I couldn't make it out until she explained it was upside down, I could see the rotor blades connected to a large ring on the floor, I also saw the significance of it being upside down, Whoever had covered it over with scrap metal didn't want the blades damaged, this could also mean there would be nothing wrong with it mechanically, The Duchess looked quite proud of herself when I thanked her for finding it.

I asked how she managed to find it amongst all the other junk, she showed me how the engines were a different colour to all the other metal around it.

= CHAPTER TWENTY ONE =

Now we had a better idea what we were looking for we soon found four more but they were all pretty well covered up in the same situation, forgetting about the cars my priority now was getting one of these flying machines to look at, asking the Duchess to collect a workforce to help to recover one of the fliers, I set to and started to clear away the debris from the first one we had found, this was going to take me a long time, every piece I picked up was interesting the weird shapes and wonderful designs fascinated me, all the metal seemed to be the same composition it must be a form of stainless steel and it was very strong, I had great difficulty in bending even the smallest pieces, the Duchess arrived with the work party and with the extra help we soon had the machine standing in a small clearing in all its glory.

Or rather its inglorious, it looked really sorry for itself, just a rickety affair of two tubular triangle shaped frames, joined by a seat with a central shaft supporting four bladed rotors two opposites had a bulge on each end and a large circular ring joining the rotor ends together, the bulges were evidently the means of powering the thing, by the appearance of the blued colouring on them they were powered by combustible fuel, although I cant imagine there being any petrol on this planet.

This was a let down of the highest order, after seeing the

capabilities of the Autobrains I was expecting something with anti gravity engines at least, instead I find it works by old fashioned ramjet engines, just like the Germans used in the flying bombs they sent to destroy London in the last war, an odd thing I noticed was they had a control handlebar this was connected to a large rudder plate, which seemed a bit on the extravagant side.

Nothing is overdone in this place, there wasn't a tail rotor this was surprising but I suppose with ramjets on the ends of the rotors there wasn't a lot of torque and the rudder plate took care of that, and the direction of flight, Something was puzzling me and I couldn't think what it was, standing back to have a look it struck me there were no swash plates or gadgets to alter the rotors without them the thing would just do a sideways somersault on start-up because the advancing blade would create more lift than the retreating blade.

This I had to investigate straight away; on close examination of the blades I found the leading edges had a slot along their length covered by a thin strip of metal which moved in and out when I pushed the blades around, now I got the idea the ramjets exhausted into the hollow rotors and pressurised them, this pressure was released on the leading edge of the retreating blade giving enough lift to compensate.

Things were beginning to get interesting again, if only I can get one flying I found the fuel tank was under the seat, of course it was empty which was a good thing, Otherwise any type of fuel would have turned into a sticky glue over time.

While turning the rotors to check the working of the slots I noticed a clicking noise now I found it was the sound of an electric spark, a small cam on the rotor head worked what must be a Pietzo crystal or a magneto coil to make a timed spark for the ignition, another large cam worked the rotor leading edge slats.

Climbing into the seat I held on to the handlebar to steady myself and nearly fell out when they moved, on examination I found they moved backwards and forwards, I couldn't see the reason for this until I'd worked them for a while, then I noticed resistance building up, Of course the Germans used compressed air to start their Ramjets, but I couldn't see an air tank anywhere, and it would need a large quantity of air to start these engines, the fuel tank wasn't big enough and the only other place was to store it was in the tubular frame itself, now I know why the air pump was located in the side frame, pulling the handlebars right back resulted in the release of air to the engines.

Getting the Duchess to work the hand pump, while I listened to each engine hiss and click in turn confirmed my findings, that the hiss was fuel being injected and the click was the spark to fire it, these initial explosions would be enough to kick start them, the problem now was where do I get fuel from, there was the remains of a still for making moonshine liquor in the cabin, as a last resort I could get it working, but I don't fancy giving the Outlaws access to strong alcohol, keeping that option in mind in case there was no other way, I decided to consult Rudo, I couldn't ask Merlin because the Homeworld would wonder why we wanted it, and would try to confiscate or worse still destroy the fliers.

When I told Rudo about finding the flying machines he was amazed he had been led to believe it was impossible for a heavier then air machine to fly, I had to admit that I hadn't seen these fly, but if I had some fuel I would fly it. How I wish I had never said those words they only caused me grief later, Explaining to him what fuel was and what it was used for, he said straight away where to get plenty, On the Homeworld a similar chemical to the petrol I described was used in the nuclear power stations as a moderator, I didn't know what a moderator was and I didn't care, all I wanted was some fuel.

In my eagerness to get the thing flying I had overlooked the main thing, who was going to fly it, no one wanted the job and being as I said I would show Rudo how it worked the job fell on me, none of the workforce had any obligations to trade with the power station in exchange for fuel, our obligations were only legal tender in this colony, I said why don't we swap one of the cars for some, everyone though this was a good idea but it had never been done before and they were a dubious. about getting permission off the high council.

"You have been swapping mined rock for food and maintenance for ages so why not swap other things, they could all see the sense in this, the Duchess and Rudo used the Autobrain to contact Merlin and put the proposal to the Homeworld we had worked out a plan to say we wanted the fuel to wash off the mess the wild dogs had left, They took such a long time to reply I expect they had to have an emergency meeting to discuss our offer, the Duchess appeared quite worried they would reject it, it was only when I pointed out trading of exchanges were already part of their society, The Outlaws traded dead dogs for food and tools, and the satellites traded rocks for food and training, so it wasn't something that was going to undermine their whole society.

She must have thought this a good argument by talking directly to the home world council, which did the trick, they answered back a little later agreeing to swap a hundred gallons of fuel for each car we could supply. This was a lot more than I expected Worra had soon picked up the rudiments of bartering, Rudo was nearly bursting with excitement which I couldn't understand until he told me this factory and the other can be adapted to turn out new cars.

"Now he said we can trade with the other colonies to get more materials and resources, they must have lots of things like Autobrains and food perhaps we could get together and make spaceships", hold on a minute I told him you cant make anything until you get some oil to refill the gearboxes on the machines or they'll seize up.

"That's not a problem he said we found a large tank full of oil on the roof it was stored up there so the heat from the sun would keep it circulating, if it had been left in the machines it would have congealed and turned to gum," I was beginning to realise how clever these ancestors must have been, they seem to have thought of everything, although I still can't reason out why they built these marvellous factories then left them so they couldn't be opened.

Why would they want to store the cars and flying machines outside, I would have put them inside the factory, then the idea came to me perhaps the factory was locked and sealed a long time before the cars were hidden, The Duchess had said she saw details of flying machines in their history, and Rudo had been led to believe that it was impossible for heavier than air machines to fly, it appears there is a conspiracy to stop any work or design on flying, I would love to get the answers off Merlin but that would alert the Homeworld to what I was doing, I even wondered if fuel was the answer, if the fliers were using the fuel that was wanted for the nuclear reactors and it was getting in short supply that would account for the ban on flying.

On Earth when petrol was rationed a lot of owners put their cars away and even hid them in case they were confiscated by the army, could the same thing have happened here, although the girls have never mentioned any wars between the planets.

I've talked to Rudo and the girls about their history but it doesn't seem to fit into any time scale, taking a rerun through the facts I've gathered so far it appears the founders that's the beginning of their existence brought them here from a distant galaxy because their resources had all been used up seeding worlds and planets on their journey through space, the Ancients were the early settlers who built the factories for some unknown reason.

Then they left this planet to carry on colonising other worlds Earth was one of the worlds they seeded at this time using different ethnic groups that had evolved on other planets, meantime their own world had fallen into degeneration, losing the will to evolve and falling prey to drugs, The Ancestors were the ones who returned and sorted out the chaos setting up the penal colonies and the obligation system, it's the Ancestors who run the Homeworld and provide spaceships to visit other worlds, they also collect the ores from here and pass some of it on to the founders.

Now I'm feeling a lot happier it may not be the correct solution but it's the best I can come up with, when I find out more facts I will put them in my history book, looking at my theory of their history, it has a parallel to our own, we had a highly civilised society in the saga of Atlantis and Mur, then we have the ancient Egyptians and the beacon pyramids, next came the Greek gods and the Centaurs, then came the dark ages with wars and strife, we could be coming to the return of our ancestors to put things right, although in our history we say it is the second coming of Christ.

Can we look forward to him setting up penal colonies on the moon and putting all our druggies and criminals in the one place they couldn't escape from Imagine the money and resources it would save us on Earth, Come to think of it there's no reason for us not doing it now, just build concentration camps in the middle of the most inaccessible deserts on Earth and supply them by air, if the inmates cause trouble stop sending the supplies.

Thinking on these lines I have a better idea that can be launched straight away. send our prison population to the impoverished nations in underdeveloped countries to be looked after and guarded by them, in exchange we pay their government for each prisoner, we would need to have welfare staff on site, to oversee the running of the prisons, but the workers would be from local areas bringing employment and stability to the country and encouraging local farmers with a ready market for their produce.

The money for this scheme would come from sale of redundant prisons for starter homes for immigrants, and by stopping overseas aid which at present is paid to foreign governments who use it to buy arms to suppress claims for democracy by its own people, however I digress this is enough of my pie in the sky ideas.

= *CHAPTER TWENTY TWO* =

There wasn't much I could do with the fliers as we called them until we had fuel, The Homeworld had promised delivery on the next collection of ore in two days time, this meant we had to get one of the cars in working condition, this could be a tall order, I had no idea how they worked and none of the workforce had any experience either, digging the cars out from under piles of scrap took half a day then collecting them together and checking to see if we could use parts off some to mend others took another half day that left us only one day to get one in working condition.

I selected one near the door and started to examine it, finding nothing out of the ordinary I got the workers to turn it over so I could look underneath, these things had a split axle with what looked like an electric motor driving each wheel small tubes led from them to a small box, then up to the handlebars, this seemed funny as the pedal to make it go, was on the floor, perhaps it went to a form of ignition switch, that seemed a reasonable assumption.

Tipping the machine back on its wheels I checked the fittings on the handlebars, the pipe I had been tracing through ended in a clip I could see where it should be connected to, so I replaced it then pressed the pedal, nothing happened, feeling disappointed I went for something to eat as I hadn't bothered about breakfast, pondering over the problem while eating, I decided to have a look at the car we came on to see if there was anything different, checking it over I saw the lead I had replaced was in the right place and everything else seemed to be in the right order.

Picking up my toolbox which I only carried from habit, because these machines had no nuts and bolts, everything was fused together Rudo used a type of gun which disintegrated the metal flux they used to assemble anything, then he just tapped them and they fell apart, I threw the toolbox onto the seat where it promptly fell off straight onto the pedal, the wheels spun like mad then got a grip, the car shot forward and before I could do anything embedded itself into a pile of scrap with the wheels screaming and giving off clouds of smoke, overcoming my shock I ran after it and grabbed my toolbox off the pedal, this cut off the power, Pressing the pedal again resulted in the wheels spinning, it looked like I had repaired the car without realising it.

The Duchess who had heard the sudden commotion and crash came running out to see what had happened, she was quite concerned about my welfare which surprised me considering the normal lack of emotions on this planet, she even helped me extricate the car from the tangle of metal wrapped around it.

Rudo and some of the workers had also come to investigate the noise, and were highly delighted to find I'd got one car working, this would take the pressure off them to get a car going, they were eager to know how I had managed it. I felt quite pleased with myself showing them how the clever Earth-man had succeeded, with a flourish I replaced the connection on their car and pressed the pedal, nothing happened Boy did I feel a stupid idiot, it had worked al-right on my car why didn't it work on this one.

Then I remembered the difference was I had gone for a meal which meant I had left the car for about an hour after connecting the lead,, When I told Rudo he said the system must be the same as the one used on the Homeworld to power portable hand tools, they worked by the sun charging a small battery, this powered a receiving unit for converting a transmitted power signal into electrical power, This explained why the cars ran without fuel and why they had no fuel gauges the power lines disconnected on the cars must be the connection to the receiving aerial, Which doubled as the handlebars that's why they are covered In plastic, the handlebars on the fliers served the same purpose the manual pump and spark generator were only for start-up after that power from the handlebar aerial took over, The designers couldn't use electrical power to drive the rotors they needed expansion gases to pressurise the retreating leading edge for increased lift, this also meant the fliers wouldn't work on Earth they would need a small generator or magneto fitted to provide a spark, although i remember the Germans didn't use any electrical equipment on their ram jet flying bombs.

Our assumption that the cars would work after being charged for a while proved correct, by the next day half the workforce were scooting around in cars they had found and dug out from under the piles of scrap, watching them playing about like school kids, I marvelled at the change in their attitude, before they had been a bit surly and uncooperative, now they had changed out of all proportion, gone was the vague doped up to the eyeballs look.

I think it's the change in their diet, once they had got over the experience of watching us eat ordinary food they had started doing the same, perhaps it was a combination of that and the fact I had been encouraging them to use their own initiative, my only drawback was I had to talk to them in sign language if Rudo or the girls wasn't around, I had managed to learn a few words of their language and knew most of them by name which pleased them immensely and it certainly helped in their cooperation.

The following morning the fuel arrived, I had expected a spaceship to land beside the factory, instead someone shouted and we all watched a cloud of dust on the horizon turn into two cars and trailers loaded up with plastic containers, the four men who had brought it just sat there on the cars until we had unloaded it, they didn't show the slightest interest in the surroundings or what we were doing, they reminded me of the first Homeworlders I had contact with, one man jumped off the second car, walked to the exchange car sat on it, then drove off following the other two, The whole episode struck me as comical, if this had taken place on Earth there would have been a lot of shouting and backslapping for such an occasion, we had opened up a new trading aspect for all these worlds, planets or whatever you like to call them.

Our work party was as eager as I was to get fuel into the flying machine, as soon as the cars disappeared into the distance they were running with cans to where we had tethered the flier to the ground, the tank was filled and we soon had workers reeling all over the place in their eagerness to prime the fuel system, by turning the rotors they had made themselves dizzy, I was automatically selected as the test pilot I felt very naked sitting there looking at the ring of expectant faces, there was no going back, it was now or never, there were only two controls the handlebars which controlled direction and the twist grip which controlled the engine speed, taking a deep breath I worked the air pump like mad and after resistance built up I pulled it right back.

Whoomp there was a muffled explosion and the whole thing jumped as the rotors shot round in a complete circle then stopped, working the pump again gave the same results, getting desperate I worked the pump as fast as I could it fired every time, it appeared the starting cycle was working but the running cycle wasn't taking over, thinking on these lines I glanced down at the handlebars connection, Bloody hell it wasn't connected.

In the rush of getting the cars working it hadn't occurred to me to reconnect the leads on the fliers pushing the lead into its terminal I undid my strap jumped down and explained to the waiting crowd what the problem was by unplugging and holding up the disconnected lead I told Worra we would resume the test in an hour, In the meantime we could get the other flying machines sorted out, fill them up and connect the leads.

There was a lot more scrap yard to search so I asked Rudo to organise the men into parties to look for more flying machines They could use the cars to drag debris out of the way and to travel about the men were happy to use the cars but seemed reluctant to start work. This puzzled me until Worra explained the men wanted to, see me flying, they still believed it was impossible.

I could understand this, they thought about flying the same as I thought about perpetual motion, "OK" I said lets have our break now that will give the fliers time to charge up, Settling down for our lunch break I thought if this was happening on Earth the men would be taking bets on this idiot breaking his neck, laughing at the idea of starting a betting school among these chaps I realised I couldn't lose, if I flew I'd have lots of obligations owing me and if I crashed and broke my neck they wouldn't be able to collect their winnings, Finishing my meal I had plenty of time to think about my flying technique I reckon leaning forward will give me forward motion and lean back to hover or land, the rudder will take care of the turns, if the engine cuts out it will be all right if I'm high enough, the wind milling rotors will let me land.

Although I saw an army helicopter come to grief at R.A.F. Middle Wallop, the pilot was practising dead engine landings but he hit the ground too hard the tail rotor broke off and the torque sent the rest careering across the airfield smashing into a hut killing the chap working there, we knew him quite well and often waved to him on his daily walk to the other side of the airfield where his job was to calibrate aircraft instruments far away from any magnetic influences.

This incident had always stuck in my mind it seemed such a shame this poor chap minding his own business working away not a care in the world, then Bang and he was no more, Putting those morbid thoughts behind me, it was no time to be thinking such gloomy things, I was about to fly a helicopter for the first time without any training and not a lot of idea how to control it.

Worra was the first to appear followed by Rudo and a crowd of workers I who were eager to see me flying, i got the impression they would have dragged me kicking and screaming into the pilots seat if I had backed out of being a test pilot.

"Come on "she said it should be charged by now."

There was no reason to put it off any longer, so feeling a bit like a condemned man on his nine o'clock appointment with the gallows I followed them out to the flier, the rest of the crowd were already waiting outside, I felt the excitement building up, I wasn't a stranger to flying having done a lot in Auster aircraft I had even applied to join a company to go whale spotting with them in the Antarctic when I was demobbed, but they decided to use helicopters instead.

Strapping myself in I realised I was in the safest place, if anything went wrong I'd be safer strapped in than being chased around the countryside by a whirling mincing machine, winding up the pressure it fired first time then stopped, trying again with more throttle it fired and carried on running, letting it run for a few minutes to warm up and for my excitement to cool down, blipping the engine I felt the machine lift, the engine was noisy but I thought once airborne it would quieten down, opening the power I felt I was moving, then realised it was because I was leaning forward, experimenting with the controls it appeared so easy moving back and forward just by moving my body weight.

I signalled for the ropes to be loosened, next minute I was looking down from twenty feet up they thought I had signalled for them to be cut, panicking I shut off the power the rotors momentum didn't slow straight away or I would have dropped like a stone the few seconds grace gave me time to open up again, and for the next few minutes I was going up and down like a yo yo until I got the balance right, I found the rest of the controls were so positive it was going to be a joy to fly, leaning forward increased forward speed but I didn't try going fast as it made me lose height, I found I could hover by leaning back but I couldn't reverse, because of the rotor drive shaft behind me.

Now I had the feel of the controls I started to do figures of eight and other different manoeuvres, then I did a few practise landings,I was looking for any problems with the handling capabilities but couldn't find any.

= CHAPTER TWENTY THREE =

Feeling as free as a bird I resisted the temptation to see how high I could go, I also wondered if the machine could do dead engine landings but decided I wasn't going to push my luck and left it for later after a few more circuits I landed, we had filled the fuel tank t capacity before we started, now I checked to see how much fuel had been used in fifteen minutes of flying, it was only two inches from the top and the tank was ten inches deep, I worked it out that a full tank would give over one hour of flying with a quarter of an hour. for reserve.

The crowd of workers had dispersed by the time I'd unstrapped myself and checked the fuel, they had been so eager to see me airborne and now they soon lost interest, Worra said it was because they thought I would want them to fly the machines, I laughed at this and said the machines were too valuable for then to damage them, but if she wanted to try I would help her, she seemed reluctant at first, but when the Duchess said she would like to try hastily adding after Worra, because she is a spaceship pilot and used to flying, this sounded logical at first then on second thoughts a space pilot can't even see the ground, although it made up Worra's mind, and it didn't undermine the Duchess, higher rank.

Tying the machine down to the ground I showed Worra how the controls worked, then how to start up, we tried that a couple of times, she was a quick learner and was very confident, next I loosened the ropes and showed her how leaning forward or back gave the hover or forward motion, this worked out very well, so I jumped down and motioned I was going to cut the rope and she would be on her own, she gave it a bit of power and shot into the air, I expect my extra weight had held it down, however she didn't panic and soon managed to balance it and before long she was copying my manoeuvres. and trying to write her name in the sky.

Glancing at the Duchess who was watching intently I could see she was eager to have a go, knowing how Worra was in awe of the Duchess and it wouldn't be right for her to show a high ranking official from the Homeworld what to do, I asked if I could teach her to fly, She jumped at the chance, in fact I had the impression she had been thinking the same, dragging one of the nearby flyers over to the anchor rope I tied it down, lifting her onto the platform seat I strapped her in, then sat beside her and showed her the controls and how to start, then how to lean for tilting and hover, I could imagine flying instructors on Earth being aghast at my methods of training, after fifteen minutes of self instruction, I was now a fully fledged helicopter flying instructor I had one pupil passed already and she was flying above us to prove it.

Actually I gave the Duchess a lot more tuition to make sure, I slackened the rope to give more rise and fall which was a bit dangerous, being tethered, too much forward motion would tilt the rotor and cause it to hit the ground, so I leaned down and cut the rope I wasn't sure if the flyer would lift two of us but with a bit more power it lifted off OK, the Duchess gave a little squeal as we slowly gained height and grabbed my arm, Which was just as well in my haste to cut the rope I hadn't tied myself in, placing my hand over hers to show her how to control the rise and fall, I noticed she didn't pull her hand away, which she would have done normally, It crossed my mind that anyone touching her would be thrown in the crushers, surely in a situation like this they would compromise and the sentence overlooked, The flying lesson went very well except it was a bit cramped with two of us on the small seat.

Seeing she was competent I told her to land and she could fly on her own, landing was a bit on the heavy side but I put that down to my weight, telling her not to fly too far and be careful I jumped off, stood back and watched her take off, seeing the two of them flying up and down enjoying themselves made me realise how things had changed out of all proportion, it occurred to me that it couldn't stay so pleasant for much longer, some one or something would come along and spoil it.

In the back of my mind I had been wondering about the people in charge on the Homeworld they must be concerned about us and what we were doing in the abandoned factories, the chaps who had brought the fuel hadn't shown the slightest interest in what we had found or what we were doing, this didn't seem right to me, On Earth we would have recruited the local people as spies or sent one with the work force, that's why the chaps delivering the fuel had shown no interest they had orders not to, which means the Homeworld know what we are doing, At a guess I would say Rudo is sending messages to the Homeworld he was the one who got the fuel for me and to think I believed him when he and the Duchess talked about home rule, maybe they think I will help more if I believe I'm helping them to achieve this.

Have I stumbled on the way these people advance by letting others do their evolving for them, in hijacking this civilization they leapfrogged too far and they can't handle it, they couldn't even open up the factories, if we were in their position and given space ships and space travel, we would only use it to dominate everyone by dropping atomic bombs on them, I expect this is the reason the space ships spying on Earth don't land and share their advanced technology with us.

If I'm being used to find out things for them I don't mind, the only worry is what happens to me when my usefulness runs out, Worra said Rip Van Winkle was taken back to Earth because he became a nuisance, I remember reading about him in school story books he was a North American fur trapper who was supposed to have fallen asleep and woke up fifty years later with vague impressions of meeting people from the future he tried to convince the locals of things he had seen in the other world but they only laughed at him and said he was mad, his only proof was looking so young but this was dismissed by the church clergy telling him they would send him back to his other world through the smoke of a bonfire if he kept telling his story, which came to us through church records.

Idly watching the girls flying around I looked to see what Rudo and the rest of the men were doing, I expected to see them inspecting the fliers they had found but after filling them up and connecting the charger they had lost interest, and gone back inside, this puzzled me compared to the enjoyment they had shown when driving the cars about, and watching the Duchess and Worra's aerial manoeuvres I thought they would have been queuing up eager to fly, Never mind their lack of interest could be to our advantage, it means we have more fuel between us.

Deciding to join the girls in the air I pulled another flier into the clearing, it started up first time and I took of without any trouble getting more confident I started doing dives to see how fast it would go, I found out the hard way getting too confidant I leaned too far forward to get more speed and overdid it, I found I was diving too steep and couldn't get back quick enough to correct it.

Panicking I opened up the engine till it screamed, I don't know if it helped or made it worse, the whirling rotors chewed off the tops of some trees which helped to slow me down, before crashing into a clump of bushes, the rotors were still turning and being dazed I wasn't sure if I was upside down, not wanting to drop into moving blades, and expecting the thing to burst into flames any moment I had to move quickly, by the time I decided the rotors had jammed and the shower of leaves settled, I saw why they fitted the ring around the rotors it certainly saved my life, if the blades had hit the branches they would have bent and sliced my head off, luckily the ring stopped this, I was smothered in chewed up branches and leaves and couldn't see anything because of the smoke, my belt had come apart in the impact and I hit the ground running fully expecting an explosion.

Looking at the scene of devastation I felt so foolish talk about fools rushing in where angels fear to tread, some workers had come running out on hearing the noise, I thought they were worried about my health but they were more interested in the damage to the flier, Worra had landed right beside me and immediately came running over, she was genuinely concerned about me, Ignoring her and the crowd I jumped into her flier which she had left running in her haste to see me and took off.

Looking back it must have looked funny to them, seeing me miss death by inches, then watching me do the same thing all over again, but it was just my old flying training taking over, If you were involved in a crash you had to fly again straight away otherwise shock would set in and you'd be too scared to fly any more.

Flying around in circles looking down at the devastation on the ground made me realise how lucky I'd been to get off with only slight scratches, catching up with the Duchess who had been flying around looking at the events on the ground, I shouted for her to land but she was enjoying herself too much to take any notice of me, so I turned back towards the factory, gradually I felt some thing wasn't quite right the machine was trying to turn me around, i felt it snatch, at first I put it down to nerves but when it nearly jolted me out of the seat I hastily headed for the first clear patch to land, pulling the connector to kill the engine there was a loud screech and for the second time that day I crashed.

The rotor had seized and the torque sent me whirling into the trees, I ended upside down hanging by my straps, anxious about fire I pulled the sleeve on my strap which opened loosing me head first into the undergrowth, which helped to break my fall.

= *CHAPTER TWENTY FOUR* =

Surveying the crash from a safe distance I saw the reason, the rotor hub was glowing red hot I couldn't see the reason for this, unless it wasn't getting any lubrication, there wasn't any oil pipes or tank fitted, I thought the main hub bearing was lubricated by the incoming fuel, Bloody hell, it suddenly came to me, we used one gallon of oil to three hundred gallon of Kerosene on jet fighters, the Americans used twice as much, this accounted for the black smoke trails they left behind, this flier had been in use most of the afternoon and had no chance to cool down.

A loud buzzing alerted me to the fact the Duchess was still flying around and her flyer could seize any minute, shouting and waving had no effect, she just waved back Then I had an idea, banking on women's curiosity getting the better of her I ran into a clearing and flopped onto the ground and listened for the noise of her machine, had she been flying around because she was too afraid to land, or was she enjoying herself too much, engrossed in my thoughts I didn't notice she had landed until she tried to turn me over, she gave a little squeal of surprise when I jumped up, she had landed because she thought I had hurt myself and was annoyed when she found I was OK, when I explained why I had tricked her into landing she changed her tune and couldn't thank me enough.

Checking her machine I found the hub quite warm, but not

dangerously hot she agreed to fly back to the factory and send a car for me and the machine, which didn't look badly damaged, I take back what I said about these machines being ramshackle contraptions, now I think they are brilliantly designed.

I'm even wondering if it would be possible to take one back to Earth with me, I think I could fix the ignition system with a battery and coil, and a small wind driven dynamo to charge it, I don't think the difference in gravity will have any effect, if the flier can lift two people here, it will lift one with ease back home, I cant see any reason for these people having any objections, surely they owe me enough, and there cant be any technical problems, they carry cattle in their spaceships, and they mentioned using flame cutting equipment from Earth on the factory doors a long time ago.

The men arrived with the car and trailer and the problem of getting the flier down was soon solved by cutting down the tree it was lodged in, I was surprised at the small amount of damage, in fact I believe I could have flown it back, even the slats along the leading edge of the rotors worked back and forth, loading it on the trailer was easy I could have loaded it on my own, arriving back at the factory my first priority was to get Rudo to show me the oil tanks on the roof, taking one of the empty fuel cans with me to fill up, I found the tanks were a lot bigger than I expected and their shape would have fooled me into thinking they were solar panels.

They were only a foot thick and the size of a tennis court they would make ideal landing places for the fliers Rudo said they were built like this for the sun to warm the oil, and the convection currents kept the oil from congealing over hundreds of years, I said it was more likely to turn it into cooking oil, he wanted t know what cooking oil was, when I told him we used it to transfer heat to our food to make it more digestible, he laughed at this and said why don't we use Ultra Sonics, or light waves to heat up food.

Not wanting to show my ignorance of the things he was telling me I explained on Earth our system worked on a different one to their Obligation codes, Ours worked on money for every hour worked we got paid in tokens which we exchanged for food or goods, if we worked many hours we got more tokens, so we could save them and buy things like cars and televisions but could never save enough to buy expensive heating machines, he seemed satisfied with my answer pondered a minute then in a puzzled voice, asked what a television was, I said it's a device very much like the Autobrains but now there is plenty of work nearly everyone has one, then he wanted to know what they wanted them for, I said we watch news items, plays, dancing and even programmes on space exploration, he shook his head shrugged his shoulders as if to say what a load of rubbish, and walked off.

His action annoyed me at first I thought he was showing an interest then I realised why would he want to know, the life here is definitely preferable to life on Earth, they have all they want, they are well fed there doesn't seem to be illnesses no rush for material possessions, there isn't any rush to get obligations in fact no one owns anything, if this place is supposed to be a sort of penal colony what's it like on the Homeworld Worra and the Duchess don't seem too keen to go back, at present I'm not sure I want to go there, I think if I went I'd never get back here, which would put paid to any chance of returning to Earth.

Walking around the oil tanks looking for a means of drawing off oil for the fliers I found pipes leading down into the factory, tracing them down I found they came out in the power house, where Rudo had drawn off oil to refill the machine gearboxes, I wondered why he hadn't told me the oil was there, then remembered how Worra Hadn't told me about the wild dogs, she said it was because I hadn't asked her, I suppose that was one of the differences, here you have to ask.

Drawing off a gallon I tried to work out how much to mix in a tank full of fuel I think the Bantam motor bikes had a ratio of fourteen to one but I wasn't sure, now there wasn't the need for secrecy I decided to use Merlin to confirm it, Worra was preparing supper so I went over to the Autobrain to check my figures, and soon forgot about them when I saw a pile of information plates under the screen, the top one was another design for a perpetual motion machine, I had told Merlin it was impossible to get something for nothing and thought he would give up the ideas, glancing at the top plate, made me stop and take a second look, my first impression was that it would work, even though I know its impossible, it looked really good on paper, or rather on metal the picture consisted of a caterpillar track stood on end in a tank of water, the track had floats hinged at intervals around it so that the downward moving floats folded back on the track, on the upward side they opened out taking them further away from the centre line upsetting the balance, causing the track to turn.

Being in two minds over whether it would work or not I sat there pondering over the problems of perpetual motion, everything has equal and opposite forces, North and South, Positive and Negative, Black and White its endless, but what if you use half of each one like using gravity and buoyancy or air pressure and magnetism, deep in thought I nearly jumped out of my skin when Worra tapped me on my shoulder, she had wondered what I was doing missing my meal and had come looking for me, gathering the plates up I followed her back to the office.

Looking at the plates again took away all my thoughts about eating, this was the real thing eagerly reading the first one it read "IN THE COUNTY OF BEDFORD JOINTLY OWNED," the next plate read "HIDDEN NEAR TO C,"I quickly sorted through the plates again to make sure I hadn't missed one, but no such luck.

Now I wondered if Merlin could only read whole words and any with Three or more letters missing he couldn't understand, tomorrow I'll ask him to include the odd words to see if I can work on them and get more of the code sorted, I gave three of the plates to Worra as they were in her language.

208 = 10 + 16 = 26 +16 = 42 + 9 = 51 = Y
195 = 15 + 10 = 25 +10 = 35 = I I
145 = 10 + 15 = 25 +15 = 40 = N N
1 = 1 + 10 = 11 + 9 = 20 = T T
94 = 13 + 1 = 14 +20 = 34 = H H
73 = 10 + 13 = 23 +34 = 57 = E E
416 = 11 + 10 = 21 +34 = 55 = C C
918 = 18 + 11 = 29 +55 = 84 + 9 = 93 = O
263 = 11 + 18 = 29 +18 = 47 = U U
28 = 10 + 11 = 21
500 = 5 + 10 = 15 + 5 = 20 = T T
538 = 16 + 15 = 31 + 20= 51 = Y Y
356 = 14 + 16 = 30 +
117 = 9 + 14 = 23 + 9 = 32= F F
136 = 10 + 9 = 19 + 9 = 28 = B B
219 = 12 + 10 = 22 + 9 = 31 = E E
27 = 9 + 12 = 21 + 9 = 30 =D D
176 = 14 + 9 = 23 + 9 = 32 = F F
130 = 4 + 14 = 18 + 14 = 32 +9 = 41 = O
10 = 1 + 4 = 5 + 4 = 9 + 9 = 18 R
460 = 10 + 1 = 11 + 1 = 12 =18 = 30 = D
25 = 7 + 10 = 17 + 10 = 27 + 9 = 36 = j
485 = 17 + 7 = 24 + 7 = 31 +36 = 67 = O
18 = 9 + 17 = 26 + 9 = 35 = I = I
436 = 13 + 9 = 22 + 9 = 31 + 9 = 40 = N
65 = 11+ 13 = 24 + 13 = 37 + 9 = 46 = T
34 = 7 + 11 = 18 + 11 = 29 + 9 = 38 = L
200 = 2 + 7 = 9 + 7 = 16 + 9 = 25 = Y

THE BEALE TREASURE CODE KEYS No TWO

200 = 2
283 = 13 + 2 = 15 = O
118 = 10 + 13 = 23 = W
320 = 5 +10
138 = 12 + 5 = 17 + 14 = 31 = E
36 = 9 + 12 = 21 + 9 = 30 = D
416 = 11 + 9 = 20 + 14 = 34 = H
280 = 10 + 11= 21 + 14 = 35 = I
15 = 6 + 10 = 16 + 14 = 30 = D
71 = 8 + 6 = 14 + 16 = 30 = D
224 = 8 + 8 = 16 +
961 = 16 + 6 = 24 + 16 = 40 = N
44 = 8 + 16 = 24 + 16 = 40 = N
16 = 7 + 8 = 15 + 16 = 31 = E
401 = 5 + 7 = 12 + 15 = 27 = A
39 = 12 + 5 = 17 + 27 = 44 = R
88 = 16 + 12 = 28 + 44 = 72 = T
61 = 7 + 16 = 23 + 44 = 67 = O
304 = 7 + 7 = 14 + 67 = 81= C
12 = 3 + 7 = 10

= CHAPTER TWENTY FIVE =

Which to me looked like the picture writing you get on Chinese stamps funny enough I hadn't seen any other signs of writing anywhere else, I must ask Worra if everyone here is taught to read and write, the girls were both busy and as it was getting late I decided to go to bed ready for an early start, Next morning it seemed to take ages for the mist to clear so I spent time checking the fliers over,. the first one I had crashed had bent and twisted the rotors totally out of shape, but its given me the opportunity to see how they are made, most other parts looked perfectly alright and could be useful for spares if they can be salvaged, there are no nuts or screws to undo its all welded together.

Where the rotors had split I could see how the exhaust gases were directed by internal ducts to the rear sweeping blade, the fuel supply came up through the rotating drive shaft, which must incorporate the fuel pump inside the fuel tank, this would also keep the fuel mixture stirred and lubricate the shaft bearings, the more I studied the fliers the more I realised how good they were, this made me think when those on the Homeworld find out about them they'll come to confiscate them, and I cant let that happen.

Looking around I couldn't see anywhere to conceal one, we had managed to find them easily enough and they had been buried and grown over, for hundreds of years, so its no good me trying to do the same, Sometimes the best place to hide anything is in the most obvious place, which for a flying machine is in the air, and the nearest to being in the air is on the roof, This presented no problem, collecting two containers of fuel, I strapped them to a flier, started it up and flew it onto the roof of the factory, I don't think anyone noticed me coming down the ladders they were too busy with the machinery.

Worra had been looking for me, I had asked her and the Duchess the night before to fly with me this morning and explore the surrounding countryside, she said the Duchess was busy with the Auto brain, they were having trouble with them on the Homeworld and she was trying to correct it, I thought this was Worra's job but I kept quiet, perhaps the Duchess was better at dealing with the Homeworld, this upset my plans a little, with my extra weight I couldn't carry as much as the girls, with a full tank and two extra gallons I flew much slower than them.

If I leaned too far forward to get more speed I lost height as I found out to my detriment the hard way, these machines were designed for the weight of the people living here, its only the fact that gravity is lower here than on Earth that allows me to fly at all, I must be two or three stone heavier than the girls, even the men don't seem any heavier it must be their diet of biscuits that keeps them all looking the same, I can understand Valda shoving her hand down my trousers to see what gender I was, I have trouble at times to make out which is male or female, It would help if they cut their hair differently.

Now the Duchess wasn't coming, I decided to stay local, it would allow us to find out more about flying, starting up and taking off went without any trouble and we were soon flying at a good height I hadn't a clue how high we were flying there were no instruments at all, I would only be able to fly as far as I could keep the factory in sight, it also means we cant fly when the morning mist is around, it would hide the factory and that's the highest thing around here, I could do with a compass providing it will work here and there are magnetic poles, I shall also want a magnet.

Engrossed in my problem of navigation I suddenly became aware of Worra flying alongside me shouting and waving to attract my attention, she was pointing to a column of dust to our left, Switching on my translator, it sounded just like the intercom when flying back on

Earth, she said it was a vehicle coming from the direction of the interrogation buildings, which meant it must be very important, we must return to the factory so we could be there to meet them, I said come on lets fly over and give them a surprise, but I got the surprise when an alarmed Worra shouted back.

"No they can stop our engines if they shoot at us," This puzzled me how can they stop an engine with a burst of electrical energy, She must have read my thoughts, as she explained, while at space training school a loaded car had run away and a section leader had fired his Tee gun at the car stopping it, but the car wouldn't work again, and had to be scrapped, by what she said the burst of energy had burnt out the power receiver, it's a pity we haven't got the means of doing this on Earth it would be useful for stopping stolen cars and other criminal activities.

Worra became very agitated and pleaded with me to return, in the end I gave in we, turned around and flew back to the factory, landing a short distance from the back doors, she seemed very upset and when I asked who the visitors were she was reluctant to tell me, I lost patience and shouted at her, that if she didn't tell me I wouldn't be able to help her, it turned out she was worried the visitors were coming to take her back to the Homeworld, her period of time on this colony had expired some time ago, telling her not to worry I would make sure she stayed if she wanted to, by telling them she was a key worker and was needed to work the master Autobrain she cheered up after this and helped me hide the flying machine.

We had about two hours at least before the visitors arrived so I thought it best to hide all of them, working on the assumption that if they didn't see them they couldn't confiscate them, finishing the cover up with plenty of time to spare, we watched the workers putting the various machines through their paces, it was marvellous to see the speed they could make different shapes and size of product it looked so simple, I borrowed my dark glasses off Rudo and watched one piece being made, the operator put the metal plate from the Autobrain into the console then left the machine, I watched eagerly as a dark shadow of the desired shape formed behind a thick glass cover, millions of tiny stars starting as a pinpoint burst within this shadow and gradually moved outwards, and before my very eyes the piece started to take form, within five minutes it had formed into a solid shape.

Watching the whole process three times, the only conclusion I came to, it was a three dimensional picture of iron dust held in suspension by a magnetic field, streams of metal particles were fired from all sides, when they collided they fused with each other and gradually built up creating a solid object, another amazing thing was when it was ejected from the machine it was barely warm, I cautiously picked it up and examined it, the outside was perfectly smooth just as if it had been ground to a fine finish, the holes looked like they had been drilled, banging the casting on the base of the machine didn't even mark it.

Rudo and the rest of the men had gone outside to meet the visitors so I joined them, the people here don't usually show such interest so it must be someone important, now I felt suspicious how did they know who was coming, perhaps it was just the fact we never had visitors, the two cars eventually pulled up in a cloud of dust, the people it contained couldn't be all that important, they were covered in it, I thought they would have had a much grander vehicle or at least been screened from the elements, the passengers alighted to be welcomed by the reception committee which consisted of Rudo the Duchess and Worra bowing to them, the other workers just looked on and stayed away from the cars either in fear or respect, but I had the impression they were not happy at being honoured by this visit.

The chief visitor suddenly caught sight of me and barked out a command to the men who had jumped off the second car and were busy slapping the dust off themselves, Now I could see uniforms appearing, Christ this must be a special occasion perhaps they are going to give me a special award of some sort, these men ran and formed a circle around me, it looks like I'm getting a guard of honour, no its not, the way one of the men jabbed me in the back to start me walking soon dispelled any misconceptions I had, Rudo had left the party and gone ahead to arrange tables and chairs in one of the offices.

As soon as we entered the room I realised we were in for another interrogation, but this time it was going to be serious, Worra and the

Duchess were sitting on my side of the table and Rudo and three visitors sat facing us, while the committee were sorting themselves out I asked Worra what it was all about, but she was afraid to speak and didn't answer, so I turned to the Duchess and asked her, she said "they were having serious trouble with the Autobrains on the Homeworld and they had traced the trouble back to us," it appears Merlin has linked to all the smaller Autobrains and they cant access them which means they cant run their obligation system which I suppose is like our bank system collapsing, it would certainly cause us a lot of trouble.

Thinking over what could have gone wrong I'm pretty sure its got to be something to do with the code word we gave Merlin, if his name isn't used before giving or asking for a programme he wont respond, keeping the code word secret as my trump card could be very difficult, I can't warn Worra because Rudo was translating for me and he could listen in to our conversation.

They started off proceedings by firing questions at Worra asking her who gave her permission to operate the Master brain we had found in the first factory and why she had used it to contact the training school, before she could answer I jumped up and said I had asked her to do it as we didn't know if it would work, and then we had to test its capabilities, This immediately swung all attention on me, this suited me, I could deal with the committee better than the girls, I carried on before they could stop me," We had no option, it was the only way to prove it worked, instead of blaming the girl you should be congratulating her on her initiative, Our team has enabled you to get this factory into production and advanced this world years ahead of where it would have been, now I had the bit between my teeth and decided to go the whole hog.

"The best thing you can do is to leave this team to carry on the good work, and to show you how good we are we will repair the Auto brains to their previous performance."

We had to wait while Rudo translated my message, the committee talked among themselves for a few minutes then gave their answer, they agreed to leave thing as they were on condition we reset the Autobrains, then they surprised me by asking if we had more cars to trade, when I said they would have to deal with Rudo with regards to how many and what to trade for, I noticed his manner change, he looked at me with relief on his face, I think he expected me to have nothing more to do with him, I realised he had to act as interpreter he was the only one with a translator.

While everyone was in a good mood I thought it would help to show our visitors around the factory, first we got the chaps to show them the machines working I asked Rudo to show them around as I didn't want them to see how we contacted Merlin and got him to alter his instructions by telling him his code was only for his personal use and was special to him, Worra managed to do all the necessary without any problem, when I let her tell the visitors she had solved their problem she really beamed, I think I've made two friends for life in the last ten minutes.

While everyone was busy I got the Duchess to get a meal ready for our visitors, We had plenty of biscuits but she surprised me by saying they would eat the same meals we were having, she said the high council often ate cooked meals with biscuits for the vitamin and anti radiation properties, before I could ask about this, the visitors all came trooping in and sat down, I had to persuade Rudo and the girls to sit with us, as they were now the official spokes men and women for this colony, at first they were subdued, and I kept the conversation going with the Duchess interpreting for me, the others soon lost their fear and were soon chattering away.

This meeting with the Homeworlders interested and puzzled me at he same time, if I met someone from another world I would want to know all about them and their way of life, but these visitors hadn't asked me any questions or shown the slightest interest where I came from, with the meeting going so friendly I was tempted t ask if it was possible for me to return to Earth then I decided to wait a bit longer, I wanted to find out more about their source of electrical power, and now Worra had told me the biscuits contained anti radiation medicine I must get hold of some to take back, now I was thinking of going home.

I realised I had a lot to do, its obvious no one will believe me when I return, in fact I wouldn't blame them, Rip Van Winkle left everything here when he went, and I cant see them letting me take anything back with me, so far the only things of interest are Tee guns Autobrains, and the fliers, I can hide a gun easily enough the Auto brain is a no go, but I might be able to buy a flier with the load of obligations owing to me.

Worra came to tell me the visitors were leaving and I was required to see them go, this suited me although the meeting had gone well for us, I had a suspicion things were going to alter, the Home world seemed to give in too easily, and the presence of the guards didn't help either, I noticed they kept apart from our workers all the time they were here, I have a feeling we shall see more of them in the future, I asked Rudo if we could make new Tee guns but he said special crystals were needed, and we wouldn't be allowed any.

= *CHAPTER TWENTY SIX* =

Regarding my return to Earth and taking things back with me I realised I hadn't much to take, I hadn't explored much of the planet, I don't know where the power station is, and I haven't even seen the mines, I'd be interested to see if their technology is any better than ours, now we have the flyers to get around on there is no reason I can't go and see them for myself, I had seen some small hills in the far distance when flying which could be spoil heaps as there were no more hills in sight, so tomorrow, they will be our destination, looking for Worra to tell her I found her at the Auto brain console, and I asked her to accompany me on a trip to the mines in the morning, she jumped at the chance, which was a relief I couldn't go on my own without an interpreter.

I don't know if she liked my company, or just enjoyed flying, she handed me a wad of metal plates from the collection plate below the screen and explained, They were the cause of the trouble with the Autobrains on the Homeworld, Merlin had delegated problem and they had spent too much time working on them.

Looking through the plates I saw they were for different types of perpetual motion machines, Merlin must have a liking for them to enlist the help of the other units, He had disobeyed my orders I had explained to him its impossible to get something for nothing, but this is frightening, if Merlin can think for himself, he could easily shut down the factories or send a spaceship off course, power corrupts, which doesn't speak well for the future, I must speak with the others and work out a plan to make sure he stays our servant, and not our master, perhaps we can threaten to switch his power off if he plays up, another idea came to me we can also use him as a hostage if the home-world tries to claim the fliers or the cars.

Worra was giving lists of instructions to Merlin, which got me thinking these people had no communications between spaceships and base, yet the Autobrains and the translators can contact each other, small power electricity is transmitted to the cars and flyers through the air and heavy power goes through thick plastic tubing, which reduces power by using smaller pipes and reducer valves, the gas in these pipes must be similar to the gas we use in fluorescent lights, as the transmission pipes are transparent and used for lighting, the only conclusion to this is that they can't transmit across the different astral planes, so how do their agents on Earth contact home, I wonder if they contact their spaceships when they are visible because that's when the ships receive beacon signals.

Sifting through the plates I found three with answers to a bit more of the Beale treasure codes. looking at the first of the plates it showed the first seven numbers of the code had been multiplied by two and this number applied to the alphabet, this resulted in Roman numerals appearing.

71 x 2 = 142 = L
194 x 2 = 388 = X
38 x 2 = 76 = X
1701 x 2 = 3402 = V
89 x 2 = 178 = V
76 x 2 = 152 = V
11 x 2 = 22 = V

Instead of cheering me up this dismayed me Merlin has already given me a decode of the first numbers which read OUR PIT TRYST, now he's saying there is another solution, no man on Earth is clever enough to encode a message using the same numbers twice, so one has to be wrong, I don’t know how to add them up, it could be a grave number or a bank code number, but never mind, I can ask Merlin for an explanation in the morning.

The second plate gave the last thirteen numbers of the code and reverted to earlier keys.

35 = 8

10 = 1 + 8 = 9 + 6 = 15 = O

2 = 2 + 1 = 3 + 1 1 = 14 = N

41 = 5 + 2 = 7 + 2 = 9 +11 = 20 = T

17 = 8 + 5 = 13 + 11 = 24

84 = 12 + 8 = 20 + 11 = 31 = E

221 = 5 + 12 = 17 + 11 = 28 = B

736 = 16 + 5 = 21 + 6 = 27 = A

826 = 16 + 16 = 32 + 6 = 38 = L

214 = 7 + 16 = 23 + 6 = 29 = C

11 = 2 + 7 = 9 + 6 = 15 = O

60 = 6 + 2 = 8 + 6 = 14 = N

760 = 13 + 6 = 19 + 6 = 25 = Y

I took this to read ON THE BALCONY, the digits of the code numbers are added together then added to the preceding code digit total, some of these have been doubled or trebled, this is consistent with other decoded parts, he got the sixes and eleven from the T = 20 or 6 and the 11 from the O = 15 or 11, if he knew he would have told me.

Tapping Worra on the shoulder I said I was going to have my supper and turn in, her eyes widened in alarm and she wanted to know what turn in meant, I said it was a term we used to mean we were going to get into bed and go to sleep, this seemed to satisfy her and she nodded acceptance, this was the first time she had ever questioned any of the translations and it made me realise, there must be many more times when the translators couldn’t find the comparative words to use, come to think of it, how would they get English in the first place, did they kidnap people from Earth and just used their brains to programme the Autobrains, if they can throw living people in the crushers for harming the people in charge there's no limit to their capabilities, this will be another lot of questions for me to ask Merlin in the morning.

At that point I must have fallen asleep, only to be rudely awakened about an hour later by a loud commotion of screams and shouts as Worra and the Duchess came dashing in panic stricken slamming the door behind them,, they were both shouting and crying I'd never seen them like this before, but I couldn't understand them until I'd turned my translator on, it appeared the wild dogs had been stalking around the factory and had suddenly attacked in force, I was annoyed at being woken up but I was more angry with myself, for not refitting the main doors, the dogs had been used to running through the factory on their hunting trips and were only too happy to attack any humans in the way.

Worra was hysterical she shouted the Duchess and some of the men had been bitten I tried to calm her down but it only made her worse, in the end I had to slap her hard across the face this did the trick and she stopped wailing, Just then Rudo came into the room looking very grave, he took in at a glance what had happened and had heard me shout at Worra not to be so silly acting like she did over a few dog bites, the look on his face and the fact he had come into our room it must be more serious than I thought.

"She's right to be alarmed." he said "it's a sentence of death anyone bitten by the dogs die within seven days, first they cant drink water, then they go mad, if they bite anyone they die as well, we have to terminate them after four days" he said there was nothing they could do, their scientists had worked on a cure for hundreds of years without success, and if they couldn't find a cure it was no good us trying, I was appalled at this callous indifference, I still cant get used to it, all through their history people had died because of the dogs, a partial solution was the introduction of the Outlaws to keep the numbers of dogs down.

Now I was the one who felt hysterical surely there was something we could do, we cant just sit back until we have to execute these people, where there's life there's hope, no way could I go back to sleep thoughts were racing through my mind Worra had calmed down, in desperation I asked her to contact Merlin, at first all he could tell me that the disease was similar to the Rabies we have on Earth.

Then he startled me by saying I must be immune because I had been bitten by a dog in the cabin and I hadn't died, the incident came back to me in a flash, Valda the girl I had saved, had thrown my bandage on the floor and used something out of a bottle on my wound then put the same stuff on her own bites I also remember the foul smelling piece of bread or something she made me swallow she also ate a piece herself, I thought it was just to show me it wasn't poisonous, that must be an antidote to rabies, if so it could be an answer to our problem, I felt it was clutching at straws but we were friendly with the Outlaws and they owed us for the dead bodies we had given them.

Everyone's attitude changed when they discovered I'd been bitten and still lived, the more I thought about it the more convinced I became that it would work, although it was our only hope, now all we had to do was find Valda's village and get the antidote, with the fliers it shouldn't be too difficult, talking my plans over with Rudo and the girls I said it may help if we took something like cooking pots or food to trade with, they all thought this a good idea until Worra pointed out if the Outlaws were short on pots they would have taken those from the cabin, Rudo had a better idea he said why not take some oil lamps they could use them,. this was a good idea especially when he said he could make me a dozen proper oil lamps in a couple of hours.

= CHAPTER TWENTY SEVEN =

I was really pleased with their efforts to help, we were now working as a team it's a pity its taken a tragedy to bring this about, getting the fliers ready, loading the supplies and making the lamps kept us busy until the early morning, we wanted to start off as soon as possible, the mist didn't usually clear until noon but I couldn't wait that long, telling Rudo to keep the factory lights on at night and a fire with plenty of smoke to guide us in the day time.

Taking off and flying up through the mist was an eerie experience, then once above the mist it was exhilarating these flyers were a delight to fly much better than our noisy helicopters, all I can hear is the swishing noise of the rotors the exhaust noise being directed into the hollow rotors cancels out most of the noise of the engines by acting as silencers, looking around it resembled a sea of cotton wool stretching in all directions, by keeping the sun dead behind us, we knew we were going in the right direction.

After an hour I signalled for Worra to fly in a circle, while I descended to find out where we were, going up and down like a yo yo created enough down draft to disperse the mist enough for me to recognise our position, I had to do this seven times before finding the cabin, landing beside it I poured some oil on the ground and lit it, the black smoke rising through the mist made a good homing beacon for Worra.

She landed right beside my machine which was a bit nerve-racking for me any damage would have finished the expedition, we wouldn't have got back in time, unloading the supplies and storing them in the back room, we made ourselves a quick meal and worked out a plan of campaign we had noted the way Valda and her companions had taken when they left us, which gave us an idea which direction to take, getting airborne and flying within sight of each other, we searched until it started getting dark landing every hour to refuel and exercise our stiff limb eventually we had stop, I didn't fancy getting lost and spending the night in the open with those dam dogs prowling about, tomorrow we can light a fire to guide us, This reminded me we had lit a fire to attract the Outlaws to the spaceship to clear out the dead bodies, it might be worth trying it again, getting Worra to stand guard I collected a load of firewood, sprinkled oil on it and soon had a terrific fire going, piling large logs on it to make sure it would keep in all night, we retired to the cabin, made ourselves supper and settled down for the night.

The flyers were so easy to fly we had tended to forget the down blast in the warm daytime it wasn't so noticeable but after some time in the cool night air it numbed the senses, and having had little sleep the night before it wasn't long before I fell asleep.

Waking up wondering where I was showed how tired Id been the night before, bright sunlight flooding the cabin showed it must be late in the morning, this didn't matter too much the mist didn't lift till later Worra was still in a deep sleep so I left her and carried on making our breakfast putting some biscuits and fruit in a box to take with us, I was surprised to hear a shout, swinging around I expected to see Worra sat up rubbing her eyes and shouting at me for letting her sleep on but she was still fast asleep, the shout had come from outside, it appeared our fire had done the trick attracting the outlaws to us and saving hours of searching.

Dashing over to Worra I shook her awake and told her the Outlaws were here she was awake instantly and would have run outside completely naked if I hadn't stopped her, although I don't think it would have mattered, they don't bother about nudity here but its better to be on the safe side, While she was dressing I went outside to greet our visitors they had come to investigate our fire and with luck Valda was with them, she came over pleased to see me and I'm glad to say she didn't greet me with the same ritual as the first time we met, In front of her party I would have turned bright red with embarrassment, at that moment Worra came running out of the cabin to greet Valda like a long lost sister and soon they were chatting away so fast my translator switched off, I had a good idea what they were saying and when Valda reached into a bag hanging around her waist and handed the contents to Worra I thought our troubles were over but this wasn't the case, she turned to me and said she only had enough medicine for two people, if we wanted any more we would have to meet with their village elders.

Worra got excited and wanted to return to the factory straight away to save the Duchess, I said don't be so callous, we have the chance to save all of them, and many more in the future, we may never get another chance like this She still argued, in the end I told her to carry on, but to take a Tee gun in case she got lost or crashed somewhere she could fight the dogs off, she soon changed her mind which was just as well, I needed her to translate for me.

Next I asked her to carry Valda for a guide, Valda seemed dubious at first she said we had flown over her village the day before, and they had covered the houses with branches and put their fires out, in case we returned, no wonder we couldn't find them, loading my flier with the lamps and oil we strapped our guide on Worra's flyer and set off, to find the village, they had left the camouflage in place, it's a good job we had a guide, or we would never have found it, we landed in a deserted village.

Valda unstrapped herself went to a piece of tin hanging up and beat it with a stick this produced a crowd of people like magic they crowded round her completely ignoring us When Valda had said you must speak with our elders she must have been joking all the people around here looked the same age then again I didn't know how old any of my friends were, Valda had disappeared. and I was at a loss to know what to do when she reappeared and led us to one of the houses I hadn't paid the buildings any attention they were still covered in branches, now I saw they were made with the came metal as the factory, they were round shaped very similar to the ones at the fruit farm, with a flat top like a toy drum tipped on its side I was intrigued by its design, but I didn't have time to examine it before coming face to face with three people standing inside the doorway, Valda spoke to these people who I assumed were the elders, I could tell by their body language it wasn't in our favour, eventually they threw up their arms and went to brush past me to leave, I caught hold of the leader and sat him down on a chair, Holding him by the shoulder, I asked Worra what the problem was, she said they wouldn't give us any more medicine because if Worra's people had the answer to dog bites they would come and hunt the Outlaws again.

He certainly had a valid argument and I must admit I was on his side for a moment it was the story of the wild west all over again, I had to think fast before this chap recovered from the shock of me grabbing him, or the other two coming to his aid, I told Worra to tell him the people who had been bitten were the engineer party from the Homeworld come to repair the generators, when they die the generators will stop the food farm will fail and there will be no more food for them, besides they owe us for all the bodies we gave them from the spaceship, tell them we will cancel their obligations if they supply us with ten lots of antidote, she relayed my message to him, he spoke briefly to his companions and they both shook their heads.

Things were beginning to get desperate but I still had one trick up my sleeve, Tell him I have brought him a present in anticipation of his help in avoiding the shut down of the power station, If he will wait here I will bring it, taking it for granted they would wait, I left the hut and collected the oil lamps and oil from the flyer, returning I set up a lamp poured in a measure of oil and lit it with my trusty fag lighter, the flame flared up illuminating the inside of the hut even though it was still daylight, by the look in their eyes I saw it had done the trick, I stood waiting for an answer and was taken by surprise, when the leader grabbed the fag-lighter out of my hand and shouted something to Worra, She said he will only give you what you want if you include it in the deal.

I was taken aback by this turn of events that lighter was precious to me it had got me out of a number of scrapes The most memorable one was saving Worra from the bull back on Earth, I hesitated before giving an answer now I knew he was keen on owning something I had a better chance of getting a better deal, Tell him I will only let him have that as its very valuable to me on condition he supplies us with ten lots of medicine now, and five for every can of oil in the future, they can collect and deliver at the food farm when they go there.

By the way he clutched the lighter I didn't need an interpreter for his answer, telling Worra to remind him if he failed to keep his word we would stop the supply of oil for the lamps the leader gave out orders and the two men left us, Worra said they had gone to collect the medicine, we unloaded the rest of the lamps and oil, As a gesture of goodwill I showed the leader how to work the lighter and told him to use it sparingly otherwise the gas would be used up in a short time, and it would stop working.

By the time we had finished the medicine had been collected and loaded, Valda came to see us off and while I had the opportunity I got Worra to ask her what it was made from, she answered they used pregnant dogs blood for the ointment and dried dog droppings mixed with special herbs for the internal use, This seemed too horrible for words but I suppose its no different to us feeding a few hen droppings to baby chickens to build up their immunity to disease, Bloody hell, it suddenly struck me, that was a lump of dog shit Valda gave me to swallow, when I saved her from the dogs, If the Duchess and the men find out what the medicine is made from there's no way they will eat it, I went to explain but she said she understood, as it was a matter of life or death she would never tell anyone.

Taking off and waving goodbye to everyone, we set off towards the sun the cabin was our first stop, to refuel and have a bite to eat, taking off a few minutes later we headed for number one factory, Rudo's men were surprised to see us when we appeared out of the blue and landed, they had seen us go past on our outward journey, we needed to use Merlin to contact the Duchess, to tell her the mission was a success, and we were at factory number one, they were overjoyed at our quick return, I think if I had been under sentence of death I would have been delirious at being reprieved, although as they say the proof of the pudding is in the eating, we cant be certain our potions will work, we didn't stay many minutes before taking off for factory two, landing there I saw they had replaced the massive doors which must have been quite a job, I even wondered why they were so big I couldn't see any reason for it.

They had also converted one of the offices into a clinic for the five patients, the Duchess had put herself in charge although she was a patient, the moment we landed she ran out to meet us she spoke briefly to Worra collected the medicine and disappeared back inside, this seemed such an anti climax to what had seemed a hopeless situation, although I'm afraid I spoke too soon, the next four days were nerve-racking, we kept two guards on the clinic door, with orders to kill at the first signs of madness.

Luckily they all survived, and actually came to see me to ask how many obligations they owed me, because this was the first time lives had been saved after dog bites, of course I refused to name a price, but Worra said I would be upsetting everyone because I would be undermining their whole obligation system, I solved the problem by saying we were even, they had incurred the injury working for me, so I repaid them by making them better, this solution was accepted by everyone concerned even though I had lost my precious fag-lighter I still felt relieved I had saved them from a horrible death.

Clearing out the makeshift clinic I hit on the idea of making it into a shower room, I could certainly do with one, the water supply was easy to install, I just bent one of the sink supply pipes complete with tap across the ceiling then made an A frame out of three lengths of guard rail and put a couple of toilet buckets for water tanks on the cross member I had intended to use it as a cold shower, but now I saw the potential I decided to make it into a hot one, this entailed me getting Rudo to help, explaining what I wanted took longer than actually making it, in less than an hour they came back with the finished article, using two of the new oil heaters, I filled the tanks and waited for them to warm up, testing the temperature now and again I was looking forward to a treat.

The only problem was I had no soap, Worra had come to watch and I asked her what they used she came back with something in a plastic bottle, I smelled like disinfectant, but it was that or nothing, I thanked her, expecting her to go away while I started to undress, instead she decided to join me in the shower, this would have been my wildest dream back on Earth, now it was actually happening to me, I still felt embarrassed, to her it seemed the most natural thing to do, and when she asked me to rub some of the contents of the bottle on her back, I threw all modesty to the wind and did as she asked, then surprising myself I asked her to do mine.

The contact between the sexes here are a revelation, brought up in a prudish Christian community to believe that looking at a nude body was the work of the devil, even telling myself I was on another world didn't make me feel any better.

We carried on showering until the water ran cold some of the workforce had come to see our project and asked Worra if they could use it, Before long there was a crowd of them jumping up and down in the cold water shouting and laughing just like a bunch of schoolboys, watching them frolicking about set me thinking now they are losing their inhibitions and were appearing more intelligent I wondered if I could get them to play football.

First I would need a football, the cow skins collected from the food farm would do for the casing and a couple of dog bladders would blow up for the inside, normally we used pig bladders but I haven't seen any pigs here, Worra couldn't help me it was quite difficult trying to explain a football to her, they had no mention of ball games in their history, even Merlin couldn't help me, in fact he made it worse by telling me I would need an air pump to inflate the footballs once we had made them.

I think I upset him by telling him if he was so clever why couldn't he design one for me and I'll get Rudo to make some, I don't think it had occurred him to do things on his own, I tend to forget he's a machine, even though he seems to get emotional at times, I apologised and told him now he's one of the family we expect him to put forward his ideas and join in making decisions, this really pleased him, I could tell by his change of attitude, he immediately suggested we get the Outlaws to make footballs for us, he said they make hats and snares out of dog skins, so why not make footballs.

I told Merlin this was a brilliant idea and thanked him for being so clever, it hadn't occurred to us to get them involved, it may even help them to get more trade with the Homeworld all I have to do now is organise a trip to the Outlaws village straight away, Worra jumped at the chance to go flying again which certainly pleased me I couldn't have managed without an interpreter, now there was only one thing bothering me, how did Merlin know about the outlaws making hats and snares from skins, or more sinister still was the fact he knew about my dog bites I've never mentioned it or had any reason to, I often get the impression we are under surveillance at times and the Homeworld seems to know what we are doing, it cant be the translators sending out signals they don't work inside buildings proved by the guards not being able to find me when I was inside the factory or the spaceship.

The idea that I'm being spied on all the time makes me feel uneasy, I suppose it's the thought it could be the Duchess or Rudo sending reports on me back to the home world, although looking back none of them knew about my dog bites back in the cabin and it couldn't be Worra she was unconscious at the time.

= *CHAPTER TWENTY EIGHT* =

Next morning we loaded up with two big cow skins, four lamps for barter and a few supplies to leave at the cabin, setting off we found a big improvement in flying conditions now I had fitted a deflector to keep the down draft off us, landing at the cabin we unloaded our supplies then took off again, we had no trouble finding the village their fires were burning and the plume of smoke was visible from a long way off.

This time there was a reception committee waiting for us, I suppose they had nothing better to do, their leader was in the front and he actually greeted us, this surprised me they don't usually show any emotions, when Worra translated it appears they were worried we had returned to collect the lamps.

I assured the leader that wasn't our intention the lamps were a gift, they couldn't understand this as no one here owned anything, Worra told them they could make something for us that would settle for the lamps, when I described what I wanted they couldn't do enough for us, soon the whole village were busy cutting and sewing the panels into a ball shape leaving a hole for the bladders to poke through, cutting the lace holes caused a problem their knives only cut gashes, we overcame that by heating the knife and boring the tip into the leather with a piece of rock underneath.

The first ball presented a problem when trying to inflate it then pushing the bladder necks inside and lacing it up, caused quite a bit of laughing and swearing eventually it was finished, standing up among the crowd that had collected around me I gave the ball a bounce, it worked a treat kicking it against the side of a house a few times it bounced off at an angle I motioned to one of the crowd to kick it back, he miss kicked it, and next minute a riot erupted as the entire village chased after the ball, it was no good asking the head man to restore order he was running with them.

Waiting until they had tired themselves out I put my fingers to my lips and gave a loud piercing whistle which I normally used on the farm dogs, but I never had the reaction I got here, everyone stopped in mid stride, falling over in shocked amazement, picking themselves up they looked around to see where the sound came from, even Worra was looking at me wide eyed.

"How did you do that do it again," she said all in the same breath, giving another piercing whistle resulted in the people crowding around me wanting a further demonstration, they had completely forgotten about the football and wanted me to show them how to whistle, soon they were puffing and blowing themselves red in the face giving me a laugh at their facial contortions, by the time a few of them had mastered the rudiments of whistling the others gathered around them for guidance, and left me alone.

This gave me a good chance to examine their houses they were made of the same plastic material as the cabin, looking inside one I marvelled at the simplicity of the design, which was the same as the storehouses at the fruit farm but smaller and they had a four inch plastic tube in the centre which I thought was for a support, but Worra told me later it was for a different reason.

I was curios and asked the villagers through Worra where did the huts come from, and why were they round, the villagers said, a long time ago their ancestors were often attacked by the wild dogs, and the huts were brought from the Homeworld for the villagers protection by a spaceship which appeared without warning, the crew bent the ship open in the middle and rolled out the discs.

Then they did exactly as I thought, they put one disk for the floor, lifted the spikes of the second disk so they were upright, then did the same with the other two disks, when they had two king like crowns with the spikes sticking up side by side, they lifted one upside down and placed it on top of the first one, the tips of the interlocking spikes went through slots in the circumference of the roof and floor disks then bent over to lock them in, that was all they could tell me but it was enough, this explanation satisfied me, it also cleared up the puzzle, of how the crates of ore were loaded in the old spaceship in the scrap yard, now I know they break in half, that means the spaceships are made in two parts and joined together later, which makes them much easier to make, I thanked Worra and said I understood now why the huts were round, to fit into the spaceship.

"No" she said "they are made round so they can move them to a different hunting area where there are more dogs, they tip the hut on edge and roll them along," I thought that was a brilliant idea, it would be just the thing for Africa, where they move their huts when following the cattle to better pastures.

After the enthusiasm shown by the villagers chasing the football it seemed funny they were more interested in learning how to whistle, by this time we had about a dozen balls to blow up and the villagers started to take an interest again, after each one was inflated I would kick it in the air to test it, then start doing a few tricks, now they were taking notice and trying to imitate me, this showed promise,, there were enough men in the village for two teams, but how can I get them interested enough to play football.

It was getting too dark to fly back so we asked the headman if we could stay, we had given away all our lamps and had to eat outside I didn't fancy sleeping on the dog skins we were offered, in the end we lay on the low bed without undressing, turned our backs to each other and I went to sleep, wondering how I could whip up some enthusiasm in these villagers to play football.

Waking up and finding Worra beside me was a shock until I remembered where I was, making the most of it I explained the rudiments of the game of football to her, don't laugh I really did, I don't think I could have done this if she had slept in her usual mode, I need her to coach the men into playing as a team, then she will have to be referee as the men wont understand me, first we have to train them to pass the ball and the rules of the game, then clear a pitch and fix goal posts.

Breakfast didn't take long I wasn't sure what I was eating so refused most of it and stuck to eating a few biscuits, getting the villagers together was no trouble, I think they wanted to please me so they could get more lamps off me, Worra told them the basic rules then we sorted them into two teams, one team played without their tunics on, lining them up I whistled and everyone dashed after the ball, at least they all stopped on the second whistle, it occurred to me they were all trying to please me, so I got Worra to take over.

Women are better organisers than men and she soon proved it, she had even managed to whistle which helped to get respect, watching a semblance of a game emerging I realised there was no eagerness to get goals, Talking to Worra at half time I mentioned this and asked if she could suggest a prize to the team who scored the most goals, pulling out my ballpoint pen to writ down the goal scorers names her eyes lit up.

"That's just the thing" she said, give it to the winning team for a prize, they have never owned anything and it will encourage them, I don't know how she explained it to them but it worked wonders, as soon as the game restarted there was a remarkable difference, now there was some fire in the game, even the villagers who had been watching out of curiosity started calling out encouragement.

Valda had returned and was showing a keen interest in the game which made me wonder, we could use her for a referee, Worra thought this was a good idea and instructed her in a referees duties, we also got her to organise different shirts for the teams and to take on the job as manager, she soon showed she was capable by getting the pitch cleared and marked out to our diagram, watching this work being done mainly by the women Worra suddenly said why not have a women's team they can play the men, this was brilliant now we can get some real competition, I had wanted to return to the factory but we spent the rest of the day instructing the women's teams.

Thinking about the impact football would have on the Outlaws as we call them it could cause a breakdown in their lives, if they spend too much time playing instead of hunting, someone's going to be upset, Worra suggested they played every seventh day like they do on Earth, I was surprised at this and asked how she knew that, she said her partner had told her he heard the cheers of the football crowds every seven days on Earth.

This sounded interesting if he told her about football he must have included information about it in his reports back to the home-world, and if that's the case Merlin will have records of the rules which will be very helpful especially as its in their own language, as soon as we get back to the factory we can get the information printed and handed out to all players.

The Outlaws were supposed to be hunters which would normally keep them fit, but I had noticed the players soon got tired and didn't appear very fit to me, I asked Worra how they managed to hunt the dogs down and kill them, She said "they fitted up traps baited with rabbit meat and just went around the traps every so often clubbing the dogs to death, at the end of the rounds they delivered the bodies to the food farm, next I asked what they did in their spare time.

"They don't have any spare time "she said, seeing the incredulous look on my face, she went on to explain, while the hunting parties are away, the rest of them go out looking for water, the dew ponds only held small amounts of water and soon dried up, this was one of the reasons they had to move their houses from time to time, this bit of news appalled me, these people are quite intelligent and I thought they would have dug out wells or made some means of catching the mist that comes every night, pointing this out they said it had never occurred to them, to do anything about it, they did things the way their ancestors had done for centuries, On the point of shouting at them for being stupid, it dawned on me, the whole set-up here had a reason behind it, making the natives spend all their spare time looking for water, kept them from causing trouble, as they say in the services a contented soldier is a trouble maker.

Thinking about it I decided I wasn't going to play god after all, I'd leave them to change things in their own way, Before we left we instructed the headman in the rules of football and gave him the title of head coach, and his two deputies as linesman left and linesman right, this made everyone happy, and kept them friendly towards us.

We loaded six footballs on each flier I don't know how many were actually made but I know it added up to a lot more than the amount of skins we had brought, by the time we eventually took off I was getting worried the light was fading, manoeuvring alongside Worra I motioned for her to switch her translator on then told her we would spend the night at the cabin, we soon found it and landed there, it didn't take long to get the fire going and soon the smell of cooking attracted every dog in the vicinity, with the fire crackling, merrily away, the cooking pot sizzling, and the forlorn howling of the dogs outside. I fell asleep dreaming about football matches between the Outlaws and Rudo's men.

Waking in the morning, I remembered my dream, the more I thought about it the more the idea appealed they didn't like each other, but that could be an advantage, the way football has turned normally peaceful crowds back on Earth into hate filled warriors has to be seen to be believed, the cheers Worra's partner heard on a Saturday afternoon could very well have been war cries.

= *CHAPTER TWENTY NINE* =

Worra must have got up early, the pots were boiling and everything ready for breakfast, it still appears funny to me the way the people here have taken to eat the same meals as me, they have spent a lifetime eating just biscuits and now they've changed to ordinary food, without any trouble, I'm surprised they still have a sense of taste, watching her busying herself around ;the place so confident in herself, was a far cry from the first days we arrived.

Then I had to more or less order her to do anything she didn't seem to have a mind of her own, looking back it must have been the effect of the drugs in the biscuits they all ate, The people who have been supplying these drugs must know by now how I've been showing the natives how to eat normal food, Surely they will want to stop me, for many years they have had a dope control over them, now I've come along and upset the apple-cart, I suppose it could be likened to the opium wars on Earth we controlled Asia by supplying opium to the natives.

Anyway that's enough of my daydreaming its time I had my breakfast and we returned to the factory there are a lot of questions I want to ask Merlin in fact I hadn't finished studying the last lot of metal plates he had given me, I hope the Duchess hasn't thrown them away, that's if she is still alive, in the excitement of getting the footballs made I had forgotten about her for all I know the serum for the dog bites may not have completely cured them.

This got me worried and I asked Worra if we could contact her, she said the only way was to land at number one factory and use Merlin, that seemed a good idea, so we finished breakfast, checked and refuelled the fliers and took off, circling the factory I was surprised no one came out to wave thinking it was perhaps because they had got used to us flying by, Landing behind the back doors, I waited for Worra to land I wanted to hide the fliers in case the authorities paid us another visit, Life had been very pleasant lately and to my way of thinking something always has to come along and spoil it, little did I realise how soon my instincts were about to crash down on me, while covering the fliers with branches to hide them something kept bothering me and I couldn't make out what it could be.

Then it hit me like a blow, there was no noise, the silence shrieked at me and I felt my hair stand on end, I had the odd feeling someone or something was watching us, Worra didn't appear to notice anything and I didn't want to frighten her, this ominous silence certainly frightened me, there should be ten men working here and they usually made a lot of noise, if they had been called away they would surely have closed the doors, if only to stop the dogs messing up the place.

Telling Worra to keep very quiet and to follow behind me, she couldn't understand why, until I pointed out the absence of any noise, she said she thought there was something wrong when no one came out to see us, we both had Tee guns drawing mine I motioned for her to do the same, creeping up to the edge of the door I lay down and peered around it, nothing stirred no lights on any of the machines, which in itself was a puzzle, it took so long to power them up, it must have been something very important to make them leave, I was beginning to lose my fear and replaced it with curiosity, there had to be a rational explanation why they had all left, that's assuming they weren't all lying dead.

The spaceship crew had all died owing to a massive release of radiation if that's the case we have no protection so we could suffer the same fate, now I hesitated, fools rush in where angels fear to tread, we had no way of knowing if there was any radiation about, although Worra did point out if there were ten dead men inside we would have smelled the bodies from a long way off, besides she said we saw them alive three days ago, the dogs would have dragged any corpses outside to eat them, She certainly calmed my fears about radiation, creeping cautiously through the factory we checked the offices and toilets without signs of anyone.

The only thing we found was Merlin was still switched on, this may have been overlooked because it faced the wall and couldn't be seen from the central alleyway, this was a piece of luck at least we could contact the other factory, Worra soon got into contact with the Autobrain at factory number two, but after a while she turned to me with a puzzled expression and said she couldn't get a reply.

Now we have a problem, why would both lots of workers leave the safety of the factories leave all their supplies behind and set off to somewhere else, or were they captured and forced to leave, this seemed the most likely explanation and the next question was where would they go, there were only three places I could think of, the food farm, the interrogation centre, and the mines, we would have seen them if they had gone to the food farm the mines were too far away my best bet was they had gone to the interrogation centre, Worra agreed with me so without more ado, we uncovered the fliers refuelled them, and set off in the direction of the centre.

We hadn't gone far when we saw a plume of dust rising from the track leading to the Interrogation Centre taking this as a good sign it showed some of them were still alive, although it didn't prove anything as to why they were travelling, the only way to find that out was to land and ask them, but this was against my instincts which were crying out for me to be cautious, signalling to Worra to follow me, I circled around the rising dust at a distance just keeping it in view, landing a safe distance in front of their advance we covered the fliers with branches then hid ourselves by the side of the track it took an hour before the party came into view, why they had taken so long soon became apparent, teams of men were pulling two three wheeled cars with ropes, on the first trailer was one of our fliers, it looked like they were taking it back as evidence.

I recognised it as the one Rudo had been working on, the second trailer held what looked like four or five bodies, the drivers and the chap next to them were strangers to me, and they were dressed in a different overall that looked more like a uniform, I remembered where I had seen them before, they were the bodyguards for the dignitaries who had visited us some time ago.

Hiding by the side of the track I was at a loss how to stage a rescue, there was no cover along the track so rushing them was out of the question, Worra said the effective range of our Tee guns was about twenty paces and only stunned people, the guards guns killed people at double this range, this news dismayed me, I felt so helpless, once they reached the centre any rescue attempt would be impossible, I had to act now, then I saw another opportunity it would be dangerous but I had no option.

Worra volunteered straight away when I explained my plan, waiting till the party had gone out of sight and sound we took off again skirting around them at a distance so they couldn't hear us then I landed Worra threw down her Tee gun to increase my fire power, while I hid by the track she flew off into the distance, she timed it spot on, as the party came level with me, she zoomed over the convoy from the rear, all the party naturally looked up in alarm.

This gave me time to dash across the open ground and shoot the driver and guard on the second car with each gun, they both grunted and as the drivers foot lifted the brakes came on automatically and even at the slow speed it was going it threw them both on the ground, not stopping I raced for the first car which was some distance away, the guard was stood up aiming his gun at Worra's flier I remember her saying they could shoot down a flier and she was well within his range, shouting to distract him, the whole scene turned into slow motion, I fired even though I knew I was too far away I hit the guard who was aiming at the flier, the shock must have numbed his arm which dropped when he fired and of course he shot the driver who had been in front of him, this chap dropped like a dead weight which turned out to be correct when we examined him later, I was loath to shoot the guard again in case It killed him.

Getting closer I saw there was no need he wasn't in any condition to do anything, he was lying semi conscious across the handlebars, taking the gun from his limp hand I noticed it was heavier and made much stronger, taking a decision which I much regretted later I threw the gun as far away in the bushes as I could, there was no sense in killing people when we had the means to just stun them.

The flier was coming overhead and I waved to let her know I was al-right, the chaps pulling the cars had all gathered round me and were trying to tell me what had happened, I couldn't understand what they were saying, only the girls and Rudo had translators, a horrible thought struck me they could be among the pile of bodies I had seen, scattering the crowd out of my way I made a dash for the second trailer, hastily turning the tangled up bodies over to check them, what a sense of relief I felt when they weren't there, Worra had landed nearby and I called her over to translate.

I was eager to hear the explanations for the tragic turn of events first I wanted t know what had happened to Rudo and the Duchess, they all started shouting at once, holding up my hand to silence them I selected, a spokesman he had seen them both escape but Rudo was limping, after that he didn't know any more, next I asked Worra to get the whole story, His story started the day after we left, ten guards had turned up at number two factory announcing that everyone was under arrest by order of the high council, they had rounded up all the workers and locked them in the power room, then they had gone to number one factory and did the same there, bringing the workers all together they demanded to know where I was, when no one knew they started to get nasty and shot one of the workers, this resulted in pandemonium as every one tried to escape, one guard died when Rudo tricked him into firing at a high voltage terminal, the electric current back tracked along the beam killing him, another guard died when Rudo and the Duchess shot him at the same time.

When both their guns overheated they had no choice but surrender, it was only the fact that the Duchess was a member of the high council that saved their lives, in the confusion of loading the flier and bodies Rudo disconnected the aerial on the first car, when it didn't start the second car crashed into it spilling the occupants to the ground, everyone tried to escape again, but the guards recovered very quickly and shot one of the workers, calling for them to surrender or get hunted down by the wild dogs and killed, well they had no option but to obey, when the numbers were checked thy found two were missing, one man told us he saw them run back inside the factory but Rudo was limping, Three guards were detailed to stop behind and capture them.

When the guards couldn't start the first car, they tied the prisoners together and made them pull both cars, that way they could control them much easier, now I had the facts, my first priority was to find Rudo and the Duchess, if they hid in the factory they would know they could be trapped, or seen if they tried to use the ladder to the roof, the spaceship shell hanging from the roof blocked the view down the central alleyway so they could have made it to the back doors and outside.

Untying the prisoners and collecting the guards guns i shared them out, it was getting late and we would have the wild dogs attacking us before reaching the safety of the factory, I was worried about Rudo and the Duchess being out without any protection against the dogs, I reasoned they would make for the cabin, my rifle was there and she knew how to use it, taking off and flying apart to cover as much ground as possible we arrived at the cabin, without seeing a sign of anyone, I landed away from it and crept up with gun in hand but there was no one there, I asked Worra to leave a note wrapped around my rifle, where the Duchess would be sure to find it, although I don't think she would shoot anyone.

This was a big disappointment I was sure this was the place they would make for, trying to think what I would do in their place, especially if I was injured, starting from the factory I would back track and hide in there, we had checked number one factory, now I realised the guards had assembled the prisoners at factory number two and started from there, and the obvious safe place is on the roof, with its one means of access.

Telling Worra I knew where they had gone I ran to my flier and was soon airborne, she followed as I turned and made for factory number one, we soon got there and after flying around it a couple of times to make sure there was no one on the roof and not wanting to land in case of a trap I set course for factory two as we approached it we saw a figure on the roof waving to attract our attention, which turned out to be the Duchess, signalling to Worra to fly high and circle to find the three guards, they must have seen our return and would be coming for us.

Seeing she understood I prepared to land flying along each side of he building to make sure there was no sign of anyone hiding, there was no sign of Rudo either, and this worried me, he may be too badly injured to climb the ladder to get on the roof, touching down right beside the Duchess I asked if she was alright,, she warned me that the high council had ordered my arrest and I must hide. I said there was no need to worry we had rescued the workers and could hold out for a long time, she was aghast at the idea of disobeying the high council, until I explained, if we remain free we can and find out what the trouble is, and we can threaten to stop the supply of ores from the mines, they need us more than we need them, if by any chance that threat didn't work we can get Merlin to close down some of their most important systems .that will certainly make them sit up and take notice, for a start we could ground all their spaceships just by overloading the system,then the satellite Abrains wouldn't have time to plot the courses.

I wonder what would happen if we asked them to work out what came first, the chicken or the egg, That's assuming they have chickens here, I haven't seen any yet.

= *CHAPTER THIRTY* =

Rudo had started out for the cabin, he was sure we would fly back into the clutches of the guards and had set out to warn us, I saw he'd made a good camp by the water tanks a supply of food and a pile of mesh to sleep on showed he was prepared to hide a long time, I didn't ask about his injuries, anyone bringing piles of mesh and supplies up vertical ladders couldn't be badly hurt.

The noise of Worra's flier circling in the distance suddenly changed and I couldn't hear it any more, then just as I started to fear the worst it started up again, shading my eyes against the setting sun I saw the flier returning with two people on board, she had found Rudo and picked him up, I knew he was terrified of flying and it tickled me to think he had no qualms about it now, with two people and supplies on board she couldn't get the height to land directly on the roof she had to fly in a circle to gradually gain height, then landed beside us.

Rudo didn't seen the worse for wear, I thought they had spent a couple of nights in the trees but he said they had doubled back and finding the guards were out looking for them had hidden on the roof, I congratulated him on such a good idea it gave a good view of the countryside and was easy to defend, I told him about the three guards sent after him, he had realised this, that's why he had left the Duchess here while he went to the cabin.

Worra had given him the spare Tee gun, which put us on level terms with the hunters, then he surprised me by saying lets go and ambush them when they come looking for us, they must have seen your fliers land, this sounded sensible to me, the sun was setting and t would be dark soon, working to Rudo's plan we descended the ladder, while he and Worra checked and locked the back door me and the Duchess did the same to the front ones, I had a feeling the guards had spent the night in number one factory, after closing the door it wasn't long before someone tried to open it, I told the Duchess to tell them if they wanted to enter they would have to surrender and throw their guns inside, otherwise I would leave them to the dogs, they threatened us with dire consequences if we didn't let them in, I told them of the dire consequences they would get from the dogs.

Now it was getting really dark and I said if I have to wait any longer I wont open the doors at all as the dogs must be on their way, telling them about the fate of the two men who had locked us in the power house did the trick, in the end they were pleading with us, opening the door six inches I waited for them to throw in their guns, they did this without any more fuss, and three guns came sliding in, knowing how crafty they can be here.

I hesitated, and told the Duchess to t ask for the other gun we knew they had, I started to close the door when another gun came clattering through, I saw this was one of the killer guns, this showed how lucky I was to pretend I knew they had a spare, this crafty move of theirs nearly made me slam the door and let the dogs teach them a lesson instead I told the Duchess to stand back and keep them covered, in case they tried any more tricks.

About to open the doors I heard the convoy returning the three guards turned and ran to meet them, jeering and laughing, thinking their mates were still in charge, imagine their surprise to find their former prisoners had reversed the roles and taken over, in the meantime we had opened the doors and they drove straight in, closing them behind the second car gave us all a sense of relief, everyone remembered how ferocious the dogs were the last time.

After taking the prisoners in and frisking them one at a time, we locked them in the power house we were at a loss to know what to do with them and the bigger problem of the dead bodies, Rudo wanted to throw then outside for the dogs finish them up, but Worra shouted to stop them, I looked up in surprise at her being so compassionate, then she spoilt it by saying the dogs will get a liking for human flesh and hunt us whenever we go outside.

Now we were safe inside the crowd looked at me for orders i felt a little embarrassed taking charge but I suppose someone has to do it and it may as well be me, looking at the condition of the men my first order was to clean themselves up and have supper, then calling Rudo and the girls into the office, I asked their opinion on what we should do, they were unanimous for us to surrender, that didn't appeal to me, especially when I asked what our punishment would be, they said we would go to the crushers and the men down the mines, that certainly made my mind up for me, when I pointed out we had nothing to lose, they agreed we fight.

Next I asked if they had any idea why we were being arrested, the way they looked at one another showed it hadn't occurred to them to find out, telling Rudo to send for two of the prisoners I explained when we find out why we are wanted it will give us something to work on, the two ex guards were brought in and it didn't take much to get them talking, twiddling one of the Tee guns in my hand soon loosened their tongues, one was a proper know all and had seen the orders, He told us the food farm had reported the outlaws delivering the last load of dogs had been wearing different coloured clothing, the guards captured two of them and got the information that we had given it to them.

A picture of revulsion flashed across my mind as I remembered the Outlaws dragging the clothes off the bloated up bodies on the spaceship, now we know what its all about we can plan accordingly, sending the prisoners back we talked over the problem while we ate our supper, I said the best thing was to contact the High Council in the morning and do a deal concerning the spaceship and its contents.

Telling Rudo to inform the men about our decision, I could almost hear the sighs of relief as we finished the meeting.

Next morning dawned just like any other morning since I'd been in this place, the morning mist came swirling in leaving the whole landscape dripping wet, if this is a man made climate whoever seems to have made it did a very good job, the temperature is just right for growing, it should be like a tropical forest, but the sparse growth must be because of the poor soil, that must be the reason the food farm needs so many bodies for fertilizer.

Thinking of bodies reminded me we still had five to get rid of, we could burn them but I don't want the Outlaws thinking it's a signal and turning up at the same time as the next lot of guards, the other solution is to bury them, I got Rudo to get a burial party to dig a large hole away from the factory and drop the bodies in, checking later i noticed they hadn't put any big stones on them, when I told the Duchess she said why bother they cant get out, I thought this was funny until I realised she was serious, "No," I said.

"We put big stones on them to stop the dogs digging them up, I don't fancy seeing half eaten bones lying about the place."

Having got the matter of the bodies settled our next business was to contact the High Council to negotiate with them, Worra soon made contact it was so quick I fancy they were on the point of contacting us she hadn't switched her translator on so I couldn't understand what she was saying but I could see she was being frightened, so I gave Merlin the password and told him to shut down, Worra was aghast with shock; she screamed at me.

"You have insulted the High Council."

Of course I replied that's the whole idea they will think we have a position of power and do what we want, if you tell me what they said we can plan our actions, she said the council ordered us to surrender at the interrogation centre for trial on with-holding information, and failing to give up equipment to the authorities, these charges were what I expected, this society was similar to the communists on Earth, working out a plan of action took me an hour which I thought was long enough to keep the Council waiting, switching Merlin on I gave Worra the messages to send, These were to tell the Council there was no way we would leave the safety of the factory, I would take them on a tour of the spaceship to show them the reason we kept it secret, and if they tried to trick us we have left instructions with the Master brain to wipe out all the records its system contained and we would destroy the remote controls for the spaceship.

Feeling pleased over the wording of the message I'm sure it would arouse their curiosity and remove any doubts over their coming, the prompt reply to say they would come in two days time aroused my suspicions its obvious they are planning to trick us, having two days before our visitors arrive, I was undecided whether to improve our defences or teach the men to play football, however thinking my situation was similar to Sir Francis Drake playing bowls before defeating the Spanish Armada and knowing the factory was impregnable I chose the football, besides it would keep the men busy.

Introducing these men to the game was harder than I expected, this lot couldn't see any sense in running after a ball and kicking it, I can't get any spirit of competition in them, the outlaws played to please me and hoping to get a lamp, but these workers never owned anything and don't want to, its no good offering prizes, anyway I got Worra and the Duchess to pick a team of ten men each, the men had all been instructed and I thought they would be more intelligent than the Outlaws but the game started off in exactly the same way, the whole two teams rushed after the ball like a pack of howling dogs, I let them carry on for a while Worra tapped me on the back and put her fingers to her mouth I got the idea, she wanted me to blow the whistle to stop the game.

Taking a deep breath and placing my fingers in just right I blew the loudest whistle I could, The chaos at the village match had been comical enough, but twenty men running at speed, suddenly stopping in mid stride, has to be seen to get the full impact, the resulting pileup was a classic, with bodies tumbling all over the place, they picked themselves up, and forgetting about their bruises clamoured for Worra to get me to show them how I made such a loud noise, now I had something they all wanted, something to fight for, Worra relayed my instructions and told the men I would teach the winning team how to whistle.

This certainly did the trick, when the game restarted it went totally different, they started to play as a team passing to each other and calling out for the ball, Rudo had made me a water clock for timing the match, after half an hour I saw they were tiring so I whistled for half time, these chaps didn't appear very fit, apart from teaching them to play we will have to start training classes as well.

While the men were resting at half time, Worra came over to me and said there was someone in the bushes waving to attract our attention, she believed it to be the Valda from the village so it must be something very important Pretending to miss kick the football into the bushes I ran after it with Worra close behind me, we hadn't gone very far when Valda suddenly appeared, she motioned for us to be quiet, then came closer and spoke to Worra,. keeping a sharp lookout in case it was a trap I waited for them to finish, by the look on Worra's face it wasn't very good news then she translated the message to me it showed the high council were planning to trick us.

Valda and a hunting party had gone to the interrogation centre to rescue the Outlaws who had been captured at the food farm, when they got there they found a large party of guards assembled, thinking they were going to attack their village they hid and listened to what was being planned, they found out the guards were to surround the factory and while we were at a meeting with the council they would take us by surprise, I was to be stunned with a hidden gun by one of the council members, Worra, the Duchess and Rudo were to go in the crushers one at a time to make me hand over the spaceship remote controllers.

Christ this sounds serious, for a start we hadn't the manpower, or enough guns, its hopeless, the guards all have killer guns, there's no way of preventing the inevitable slaughter, we don't stand a chance, for the first time since arriving here I felt despair, I've got these people into this mess and I cant think of a way out, it even crossed my mind to involve the Outlaws but that would only add to the dead, I tried to thank Valda for bringing us such important information, but she said she still owed me many obligations for saving her from the dogs.

Returning to restart the match my heart wasn't in it and I was glad when it finished. I went to bed that night a very worried man, lying in bed I thought of a plan that was better than nothing, if I held the meeting in an office with an intact fire extinguisher I could let it off at the meeting I'm sure the stun guns won't work when soaking wet, I must remember to ask Worra about it in the morning, all we could do now was to wait for the arrival of our visitors in two days time, sleep must have overcome me at this point as the next thing I remember was Worra waking me, "Come on" she said the men are waiting for you, I was still half asleep and didn't understand what she was talking about, she said.

"The men have been playing football since first light but they need you to organise the game, they get excited and are even hitting each other, we never have any trouble when you are in charge, me and the Duchess tried to keep order but it didn't work,"

I had to laugh Worra was even calling her friend the Duchess, having a hasty wash and breakfast I trotted out on to the pitch, and noticed straight away it had been marked out and levelled, there were even two goal posts with a cross bar, this puzzled me I hadn't told these people or the Outlaws about goal posts, I was about to ask Worra when the men called out for me to start the game, and they had called out in English, this didn't surprise me as much as I it should have done, all the countries on Earth use the English words associated with football.

Worra carried on with her duties as linesman, with one of the chaps on the other side, how she had managed to instruct him in the fine arts of the game beats me, it seems so unlike these people to be interested in anything, a few days ago I would have said it was impossible to teach them, but now they can't learn fast enough, looking around I noticed the Duchess sat by the side of the pitch reading from a metal plate, waiting till half time I went over to see her, she showed me the plate which contained a plan of a football field, with lines of their writing on it, I was amazed how could Merlin have acquired such detailed information.

The Duchess was the obvious person to take over the job of referee, the chaps all respected her and she carried authority, we have no women referees on Earth but there's always a first time, seeing her with the plates reminded me a haven't looked at the last ones from Merlin, if I get her to take over this game I'll have time to check them out, she didn't need persuading. In fact she jumped at the chance I think the men were disappointed however she soon took control of the game, she was a natural referee, after the final whistle, which brings me another story.

Rudo had got Merlin to do the design and he made the whistles, but I had to confiscate them, as all the people watching the match were continually blowing theirs upsetting the game,

I congratulated the men on a good game then singled out the girls and praised their efforts, especially in getting the instruction sheets off Merlin, whose old records certainly improved the game. now everyone knows the rules.

= *CHAPTER THIRTY ONE* =

Next I asked Worra if she could get me more instruction sheets off Merlin to give the Outlaws providing they could read, "Yes" she said"its compulsory to read at an early age, then she asked why I wanted to teach them, I said, in the future we may be able to arrange a match between your team and theirs, she stepped back in amazement, that would never be allowed she said, we can't have any contact with the Outlaws, our ancestors used to hunt them to keep their numbers down, that was until they stopped breeding, this was news to me, but she wouldn't or couldn't tell me any more about it, so I let the matter drop.

Waiting until the next game started I made my way back to our living quarters to check the last lot of plates from Merlin, there were five altogether three were perpetual motion ideas and two were decodes, taking these first I was disappointed, expecting to find a complete decode all they contained were bits and pieces some parts made sense, looking at the other plates I found I couldn't concentrate.

I had more pressing things on my mind, knowing how crafty and callous the High council could be I had doubts over our plans, this brought me to thinking about the fire-power advantage they had over us if I could find a way to shield ourselves we would gain the upper hand, In the middle ages on Earth they had metal armour for protection that would cancel out any electrical charges, the problem is we only have about two hours before they get here and there's no way to make enough suits in that time.

The only other material that may be of use is the mesh we used to make the lamps fireproof, using two sheets overlapping to reduce the size of the holes and leaving a strip to touch the floor which would earth the electrical charge when fired at, this all sounded too easy but it was worth a try Finding Rudo on the roof I put the idea to him, he was enthusiastic and pointed out the mesh was conductive otherwise it wouldn't work in the plating process, making the suits was easy, we just cut off a length and cut a hole for the head to poke through, after making the first one, we asked for a volunteer to test it, Rudo laughed and said would I try it, I must admit I had no intention of getting shot, the memory of being shot in the interrogation room would stay with me for ever, Rudo solved the problem by using one of the prisoners, they didn't bother to tell him what he was doing they fitted him up with the suit and fired a Tee gun at him, the shot struck him in the back but he didn't even flinch, four more shots were fired with the same results.

Then they brought out the killer gun and gave it to me to try out, there was no way I could shoot a man in cold blood, Rudo had no qualms about it noticing I hesitated he took the gun off me, pointed it at the prisoner and fired, he yelped and collapsed in a heap, Rudo threw the gun down on the floor in disappointment, then kicked the body in temper, I was more annoyed at this callous indifference to killing an unarmed man than I was in the failure of our experiment, bending down to retrieve the gun I noticed the body was breathing, I rolled it on to its back and was about to give the kiss of life when it groaned and sat up.

I asked if he felt al-right, this produced no response, until Rudo shouted at him and relayed the answer to me, he said he felt al=right but wanted to know why he was still alive, after being shot with a killer gun, I couldn't help laughing this chap had seen the gun aimed at him and fainted when it fired, which to me was a natural reaction, seeing how frightened the poor chap had been, I asked Rudo to set him free as a reward for assisting in our experiments.

After blowing the whistle on the football match we demonstrated the body armour to the men, surprisingly enough the former prisoner volunteered his services as the target, he seemed quite proud in his new position as the focus of attention, seeing they would be invincible in their new body armour, the men set to and were soon parading around in their own creations which wasn't too soon, a shout from the roof informed us the convoy carrying the Council members was approaching the factory, contrary to their promise of only a few guards for protection they had an escort of guards strung out on each side forming a moving barrier, I suppose this was to guard against ambushes.

This show of force made our plans useless, quickly getting the men into their armour I positioned them inside the doors, however the convoy must have suspected a trap, they stopped just outside,, The master of ceremonies, well that's what he looked like, held up a flag and gave a speech, which I couldn't understand until Worra translated for me, he was appealing to the workers to hand the leaders over to them, in exchange for a free pardon.

Stepping forward to address the leaders I was immediately fired at by one of them with a hidden gun this was a signal for the guards to attack, they charged en mass into the factory where our men met them with a concentrated fire, not expecting such a hot reception they went down like ninepins, the ones who managed to fire back were bewildered by our men not falling when shot, our shortage of guns were nearly our undoing, until our reserves took a hand engaging the guards in unarmed combat.

I led the way when my gun overheated my extra height was a big advantage a guard fired at me and expecting me to fall brushed past, I promptly punched him on the head and he went down, the chap following him I hit on the chin and he crumpled up, this was getting exciting, the fighting was developing into hand to hand combat, with individual fights taking place all over the factory, seeing we were gaining the upper hand I thought I could shorten the fight by grabbing one of the councillors to make him call for surrender, dashing outside I found they had all disappeared into the trees noticing some of the guards shot in the first attack were recovering I shouted for them to be tied up and made prisoners.

We had fewer guns to start with and the guards were fitter and stronger, but we had the advantage of our primitive armour which had so bewildered them, although our mesh capes only partially covered us, it was surprising only one of our men had been killed, while the guards were like sitting ducks, another bonus came to us when we collected the unconscious prisoners, a number of Tee guns were found which hadn't overheated, these were used to full effect to subdue the last of the guards holding out in one of the offices.

Now we had won the fight it hasn't solved our problems, we have a load of prisoners to feed and look after Rudo wanted to hand them over to the Outlaws but I overruled him and said it was in our interest to keep on good terms with the authorities our best plan was to let them free, and to find the runaway councillors to seek a settlement with the Homeworld, we still have the bargaining power of stopping the shipment of ore from the mines.

The prisoners didn't seem very keen to be set free, Worra explained, as it was getting late they would have to spend the night in the trees because of the dogs then when they return to base they will be sent to work in the mines for failing in their duty, this seemed a bit harsh but its better than going into the crushers.

After organising a burial party and clearing up we split into pairs to find the missing Councillors pairing up with the Duchess we set off, expecting our quarry to have only gone a short way, I wasn't unduly surprised when we caught up with some of our former prisoners, even when they surrounded us I wasn't alarmed I had my Tee gun in my hand and the Duchess had hers, the crowd parted and one of the missing Councillors stepped forward holding a killer gun, he spoke and I didn't need a translator, handing over my gun, I felt terrible to be captured so easy, after all the planning and even winning the fight made me feel such a complete idiot.

The Duchess was talking to the councillor, she translated and said we were waiting for the others to be found, eventually they turned up and we started off back to the interrogation centre I asked her why we hadn't got the cars, she told me, in the fighting they had been hit by many shots and rendered useless.

It began to get dark and I was thinking of ways to escape, but there was no chance two guards were detailed to watch us all the time, when we stopped to camp for the night we were tied together this made sleeping impossible and going to the toilet very embarrassing, although it didn't bother the Duchess.

All night long the dogs sniffed and prowled around outside the barricade of tree branches the men had built, occasionally one would scrabble through, the men would shoot it and throw the body back over the fence, this led to terrific fights as the other dogs ripped it to pieces this howling and barking attracted dogs from miles around and by morning there were hundreds around the camp as it got lighter they drifted off and by full daylight they had all gone.

After striking camp it took about four hours to reach our destination, expecting to get a meal I was annoyed at being shoved into the showers, luckily for me they had untied us, I still cant get used to the idea of no discrimination between the sexes, then we were escorted into a room where the reception Committee was waiting for us.

= *CHAPTER THIRTY TWO* =

And what a committee I nearly burst out laughing, if these Homeworlders were trying to impress me, they certainly succeeded, for the life of me I cant see where they got the idea from, four Indian Gurus were sitting cross legged on raised cushions behind a long table, dressed in traditional Indian clothes except they had no turbans on, all had the regulation waist length white hair, looking at them, I immediately thought of Merlin, that's it, these people must think I often consult Merlin the Autobrain when I have problems, and to them it looks like I respect him, so they've adopted this getup expecting me to respect them.

What's puzzling me now is why or what are they going to question me about, they know all about me and the people on Earth so why all this play acting, it doesn't make sense, they have no compunction about throwing live people into the crushers, so why go to all the fuss of a sham trial, unless its to be a propaganda trial like the communists stage to remind the peasants whose in charge, I cant think of any other reason.

No one had spoken yet so to start the meeting off I gave each of the Gurus a proper Sir Walter Raleigh bow, sweeping my hand from my waist outwards and bending right down low, I meant it as a sarcastic gesture, but they took it as a compliment, and I noticed their stern expressions soften a little, number one Guru spoke in his own language, I couldn't follow what he was saying, when I didn't answer they realised the trouble, and one of the guards was sent to collect the Duchess to translate, it was some time before they returned which gave me time to look around.

There were no other guards or attendants in the room, just two Homeworlders each side of the Gurus I could forgive them for not having an interpreter, but to have no guards was beyond me, they must know how I kidnapped the Duchess at an interview like this one, they hadn't even searched me I could have a hidden gun.

The messenger sat down and the Duchess stood by me, number one spoke again and she switched off her translator which meant I only understood bits of their conversation, eventually she turned it back on and I was included in the proceedings.

The Founders, said they had been monitoring me since I came here and were interested in my motives for trying to improve the lives of these people, who had been perfectly content to live the life they had evolved to, in my short time here I had awoken the same spirit they had seen on Earth, the desire to advance, to better ourselves, and enjoy life, He said they had striven for many generations to instil those qualities in the people here and on the Homeworld without success, and had consequently given up on this outpost and bypassed it on their colonization quest in space.

Automatic progress messages of events here, had alerted the Founders and aroused their curiosity, my progress in educating the people was very similar to the methods they used in starting up fledgling colonies, they left an advisor to instruct the inhabitants in husbandry and agriculture, if they left more people it caused resentment and they were often attacked and killed, this had happened many times through the ages.

Earth had also been bypassed, and left to sort itself out, infrequent progress reports of our constant fighting and killing had kept the Colonists from returning, then when I suddenly turned up and the messages received were contrary to their understanding of us, it aroused their curiosity, and they had come to see for themselves.

Their knowledge of us wasn't very flattering, they mentioned our love of war, the way we enjoy killing animals and the way we oppress other people, I asked what they meant, by that, he answered "your country has fought all over your world spreading death and destruction, enslaved half the people, and took their lands, when others tried to get their share of the plunder you turned to fighting each other often using the conquered natives."

I felt I had to interfere at this stage and told him his information was out of date, things had changed and they should look at the plus side, in the distant past we had been warlike, but over time we have evolved peaceably, in the countries we acquired we abolished slavery, educated the native peoples, invested money, built roads and railways, established rubber plants in Malaya, tea in India, sugar and countless benefits in all parts of our Empire, so they could trade with us and gain self respect, in time we handed their country back to them and helped lawful government to power, when our neighbours made war on us these former oppressed people who according to you should have despised us and glad to be free, rushed to our aid, and fought by our side time and time again.

Number one held up his hand to silence me, then ignored me and engaged in earnest conversation with the others which took a while before they reached an agreement, the leader addressed me through the Duchess.

"We are impressed by your conduct "since arriving here you have given us much thought, we never understood how your people could attack an opponent then tend their wounds and nurse them back to health, and to make it more confusing you would make friends and even intermarry with them, we have spent much time to analyse your motives in saving the life of Worra the spaceship captain, then you saved the Duchess twice, you tried to save the workman electrocuted in the pit when you knew it was hopeless to revive him.

We cant forget the two men who tried to kill you and you spoke up to save them, you did all those things without thought of personal gain and at considerable risk to yourself, you owed no obligations to any of these people and by your own nature you would do it all again, without expecting any favours or reward.

Through you we have a clearer insight into factors that influence decisions regarding Earth, our agent reports have been absent over recent years we suspect damage to our beacons which have made communications difficult, and your flying machines chase our ships when they appear to pick up signals, Reports from our agents, this remark hit me like a brick and the other bit about influencing our decisions regarding Earth shook me to the core These people are saying they have spies on Earth and they can influence our lives.

Bloody Hell some more bits of puzzle fell in to place, Worras partner was one of their agents and he had been masquerading as old Tom the odd job man on the farm, this would explain his long absences when he was supposed to be visiting friends in Scotland and It also explains who owned the box of toiletries I managed to steal off the spaceship, when It landed here.

I felt quite relieved to have fitted some more bits of my puzzle together, it crossed my mind to sneer at their idea they could control events on Earth, then again I wouldn't gain anything by antagonizing them, looking back it seemed they were on the point of asking me a favour, otherwise why would they be telling me these things surely they would want to keep them secret, the sudden jolt of the revelations had distracted me, and it was a while before I realised they were still talking to me.

Football that was the word that brought me back to reality, the leader was saying the idea of kicking a ball about has awakened a spirit of competition in the people here that we thought had died out many decades ago, we want you to instruct others in this pastime to get the people of this planet to think for themselves again.

At this point I couldn't help myself by saying if you influence the lives of the people here and you want them to be independent, why have you been feeding them drugs in their food supply all these years, and why did you send guards to kill me, and to enslave the workers.

The leader held up his hand to silence me, then turned to the Homeworlders, I couldn't understand what they were saying but by the general attitude I knew they were being interrogated, later the Duchess told me the law to dope the food had been used many ages ago to suppress the carnal natures of the original colonists, and it had worked so well the leaders had kept it, without informing the Founders, non drugged food was allowed for the council and partly doped food for the technical staff, this form of people control had been going for so long it had been accepted as normal, any dissenters who realised what was going on were outlawed and banished to this colony, they must have been the original Outlaws.

The committee had been talking amongst themselves while the Duchess was talking to me now the leader turned and said, a great injustice had been done to these people, we believed their backward evolution was because they were lazy,. but you have changed our opinion, we have also decided to reconsider our attitude to Earth, if others of your race are as considerate as you there is hope and we will help you.

At this point I had to step back and pinch myself, I couldn't make up such rubbish, even in a dream, biting my lip till I tasted blood I still couldn't believe I was awake, In desperation I pinched the Duchess so hard it made her squeal and jump in alarm, this didn't change anything the scene before me stayed the same, accepting the facts I decided to play along with any suggestion they made, almost as if they could read my thoughts number one asked if I would help to get the population off their dependence on drugs this seemed an easy request so I agreed, it would be easy enough to destroy the plants, they had already mentioned football and now they asked if there was anything I needed to start a school, this request really flipped my mind and I told them if they can travel through space and time surely a small planet like this, would be only child's play for them to put right.

The whole committee looked at each other in embarrassment, finally the spokesman said "we are not allowed to interfere with evolution anywhere in the galaxy, we can only encourage and guide, in the case of these people it would take too long they will lose their place in the order of levels."

This was the first time I had heard any mention of them, and I was interested, if they want my help I want something in return, I can't ask them straight out it may upset their opinion of me, so I said nonchalantly, we know about levels on Earth, but we wonder how entities cross from one level to another, over the ages we have had jumping Jack Flash, Germaine, The Postman, religious visions, like Zebidiah ghosts, apparitions, and hundreds of other things like black dogs and poltergeist activity, .we get that many we get annoyed and would like your guidance to help us to get rid of them.

Number three answered my question, he hadn't spoken before and I had the impression he was the leader although he hadn't taken any part in the previous discussions, he said, "the entities or apparitions you are experiencing are the souls of people in torment between states who had done evil deeds on their level to themselves and others, their own conscience puts itself on trial and gives the verdict, if its down levels the soul resists and is in torment, rising to a higher level the soul accepts its transformation like a caterpillar changing into a butterfly, and flying free.

"Wait a minute," I interrupted ",your saying if I was born to cannibal parents, killed and ate other people, I would be dammed for ever, "no" he said "that's like armies fighting a common enemy, your conscience joins with others and believe its for good, the evil ones are the leaders who corrupt others with false promises to kill themselves, and as many others as possible, life is the ultimate gift to man, to take it or to end it yourself is throwing the gift back into the redeemers face."

If these people are trying to scramble my brain they're making a good job of it, I don't know what to believe, it all seems so silly why are they pretending to be gods, they know I understand there's no god, we are like atoms in a solid all contributing our little bit. to the common good, and they want me to start a football team to propel these people into their rightful place in the universe, pull the other leg its got bells on.

I don't seem to be getting anywhere in getting important questions answered in exchange for starting a football school, I had only been with the Outlaws a couple of days and they had learned very quickly, Rudo's men had also picked up the game in a short time, it would be interesting to get them to play each other especially if the Outlaws were given free pardons, maybe I could get one for myself and a free trip back home, Can you imagine me arriving back on Earth and telling the authorities I had been setting up football games on another planet, I'd be certified straight away and carted off to the funny farm.

The thought of going home made me feel homesick, this hadn't happened before and it made me realise I've been so happy here, the weather is perfect, there is plenty of food, I get on well with the inhabitants, and I still have a lot more places to visit, in fact I'm having the time of my life, so why am I feeling this way, perhaps its because everything seems too good to last.

I was brought back to reality by the Duchess pinching my ass really hard it made me jump, They want an answer she hissed, I must have looked confused because she repeated the question, I apologised and had to admit my mind was elsewhere, they had been asking what equipment I needed, I started to give them a list them changed my mind and said I don't want anything, they were puzzled at this and I explained if we make everything ourselves a lot more people will be involved and they will get interested, this made me think we need supporters and they are all on the Homeworld, I wondered if we could use the Autobrains as television links, as soon as I mentioned this the Gurus showed an interest in the idea, there are many units in the factory where they made them which we could have, that took care of the viewing part, but when I enquired about cameras they looked puzzled, the Duchess came to our aid by explaining the viewing units were also capable of sending pictures.

I should have known this, when we first contacted Worras training school we had received pictures of her friends who were still there, when the council asked if there was anything else, I wanted I thought now was the time to ask a few questions concerning them, to keep on a friendly footing I asked their permission first, of course they had to agree, I said I had noticed throughout the meeting they hadn't eaten or drank anything, were they here in spirit or physical form.

The leader laughed and pushed his hand through the table as an answer, Now I started to think of all the questions mankind would like an answer to, first I wanted to know where he came from, he must have misunderstood me, he said they had evolved over millions of years, I took a deep breath and asked him do we reincarnate as another person in another lifetime, he looked around at the rest of the committee as if to get their support, they didn't seem interested in the proceedings so he answered.

"No" he said you return as you, through the double helix, your spirit evolves from the level earned in your previous lifetime, he went on to explain, have you not wondered why some primary humans are born with advanced skills, I don't know what a helix was, but I had often wondered about people like Mozart being child prodigies' I also liked the bit about primary humans.

= CHAPTER THIRTY THREE =

Did he mean babies or Monkeys looking at the Duchess to ask what he meant I saw her eyes widen in surprise. Turning to see what had alarmed her, I had to blink twice, the four Gurus had disappeared leaving the Homeworlders on their own I don't think they noticed their companions leave, I felt cheated, had they left to save answering any more of my questions or were they recalled.

We all looked at one another at a loss what to do I had been expecting to fight for my life instead I had more or less been given control of the planet, none of this makes any sense when I asked the Duchess how we could contact the Gurus she said it was never allowed, only a full meeting of the council could summon them and only on a dire emergency.

This news alarmed me, "you mean, they were sent for without our knowledge, and what was the excuse for an emergency call out," she ignored my question and carried on," the last time they were called they decided to send the other Earth-man home, now they want you to instruct us in the game you call football," this is a funny world with their code of obligations they could have thrown him in the crushers, instead of taking him back, come to think of it if he went back through causing trouble, I should be going back with full military honours, although if I remember correctly, they were fifty years out in their timing, I wonder if they did that on purpose, knowing no one would believe his stories.

This aspect of returning home put a damper on my expectations, fancy getting back and finding you were fifty years older, would be like losing half your life time, thinking back to Rip Van Winkles story he was supposed to have disappeared for fifty years then reappeared looking the same age as when he'd left, not wishing to dwell on these unpleasant thoughts of being dumped back on Earth I decided to get on with the football programme.

First of all I'm going to make a list of all the questions to ask the Gurus next time I see them, such as how does gravity work, how can we combat cancer, or how can we live longer, how can we overcome radiation sickness, can we speak to the dead, space travel where does our soul live, who was Jesus, Atlantis, the list is endless, the trouble is answers only bring on more questions perhaps it would be best to leave well alone, and get on with the jobs in hand, I still have plenty of things to do.

With the meeting finished the Homeworlders jumped in their car and disappeared in a cloud of dust, about a dozen guards had turned up and were wandering about the place I ordered them to clear a patch of ground big enough for a football pitch, which would keep them busy for a while there were also quite a few staff who managed the place, I sent the Duchess to look for any form of transport, while I looked for anything that would be useful to us.

All I found was a pile of food containers which I assumed was ready for despatch to the mines, the Homeworld councillors must have commandeered the transport to come out to the factory, now the cars were back there and we were stuck here, then the Duchess surprised me by saying, why not contact Worra she can send the cars and we can take some of the food back with us.

This was a good idea but I was concerned about the miners they would have to starve until the next delivery, while looking around the place I had seen hills in the distance that looked man made, when I asked about them she said they were the spoil heaps from the mines.

Thinking about the plight of the miners, without their supplies, I reckoned if I turned up with a load of food i would be welcomed, now I put the plan into action, The Duchess had already contacted Worra on her translator and asked her to send two cars she said there was only one available, I asked her to bring some footballs and instruction plates, the Duchess knew what I had in mind and asked if she could be referee, she looked disappointed when I told her I would take the first game, as I wanted her and Worra to captain each team, at first I thought the guards could play the staff then decided to mix them up.

While waiting for Worra to arrive I got the men together and asked the Duchess to explain the game to them, I thought we would have an impossible task, but the Duchess commanded respect and the men were keen to learn, I think the guards were grateful to me for saving them from working in the mines, while this was going on, Worra appeared driving the one car on her own, I shouted at her for being so foolish, she could have been waylaid by dogs or some of our ex prisoners who were still wandering about, She seemed upset at my concern, and looked at me in a funny way, I hastily added that she was a valuable member of our team and I didn't want to lose her, that put her at ease, but it unsettled me a bit, I still can't understand these people, any show of affection frightens them.

The idea of going to the mines pleased both the girls Worra had visited them as part of her training and was looking forward to seeing them again, I had to curb their enthusiasm by telling them we had a game of football to play first, they accepted this with grace and we started to get one organised, the people here all looked the same to me, but the way the two captains picked their players looking for the fittest, made me realise the will to win was already present.

The captains had given them a briefing and they kicked off, the girls must have explained the game well because the game was nothing like the games we had played before, there was no mad dash after the ball, this was organised disciplined play, to think the first time a ball had been kicked on this planet was when I kicked the shrivelled head of the mummy along the deck of the spaceship and out through the door, the game had progressed so far in such a short time made me wonder, each time we had introduced a group to playing it had got easier, its as if they are all in contact with one another, perhaps Rudo has already established a link with the Autobrains, if this is the case I can't see the authorities letting me teach the miners to play football, it will upset their production targets, the staff told us it would take quarter of a day to get to the mines so we decided to start off early in the morning, this gave us plenty of time to play a few more games before dark.

Looking around I saw we had a few spectators and they had even been shouting every time a goal was scored, the match ended in a score of five one, this seemed a bit lop sided to me until I counted up the players, and found fourteen without shirts and eight with shirts, I don't think this was deliberate, but I pointed out to the girls it was their responsibility, the problem of linesmen and referees worried me, but it was soon solved when the Duchess gave the overseers at the centre Merlin's plates and instructed them on taking charge of the game, they were really pleased at their promotion, I'd noticed before the pleasure these people get from being given authority.

Restarting the game which ended early as the light was fading I picked up the ball and was immediately surrounded by players shouting at me, with them all shouting at once I couldn't make out what they wanted until Worra interpreted and said they wanted to know if they could keep the ball, they knew we were going to the mines in the morning and wanted to carry on playing while we were away.

Lying in bed half asleep and thinking of the events of the day I wondered if I was already asleep and dreaming, I mean I've never been interested in football and don't know many of the rules yet here I am organising a whole planet to play sports, what would happen if I introduced cricket golf, and all the hundreds of sports we have on Earth, we could end up with a sports crazy planet, unless the authorities put a stop to it, and its obvious they will have to do something otherwise the planet could grind to a halt, I have an idea that may save the situation, they can adopt our system and use the sixth day of the week for sport and the seventh as a rest day, I put this proposal to the Duchess and she contacted the Homeworld to ask if it could be adopted.

Their reply was totally unexpected, Three of the high council had suddenly died and nothing could be done until their successors were installed, I told the Duchess this seemed a bit of a coincidence three of them dying at the same time she said it had also happened after a Founders previous visit, this sounded a bit ominous to me but there was nothing I could do about it.

Next morning we loaded up and started out for the mines as soon as the mist cleared, the Duchess stayed behind to supervise the football games, she had appeared keen to visit the mines the day before, but for some reason she changed her mind, in one way it suited me better as it meant we could carry more supplies, owing to the extra weight we were carrying the journey took longer than we expected, and we were surprised when approaching the mines the men came running out to meet us, and running alongside us and started to unload as soon as we stopped.

I asked Worra what all the rush was about, she caught hold of a worker spoke to him for a few minutes, them turned to me and explained, they had no food for three days and the men were refusing to work, guards had been sent for but they hadn't arrived, I didn't have the heart to tell them they were busy playing football, now we had brought enough food to last a few weeks, the danger was over and everyone breathed a sigh of relief, we couldn't have timed it better, everyone was grateful and couldn't do enough for us.

My request to look down the mine resulted in a crowd of miners surrounding us eager to show us around, first they fitted us with a suit very similar to the ones I found on the spaceship, the same type air bottles were in use with a pipe leading up to the face visor, then we were given a drink, I asked Worra if they were going to give us any tablets to counteract radiation sickness, she said the drink was for that, so much for my plan to get some to take back with me.

Boarding one of the tractor units which ran on rails I was immediately intrigued by the string of trailers it was towing they were three wheelers but the front wheel was a large ball I couldn't see any sense in this it wasn't even touching the track, with two of the men as guides we travelled on a slightly descending smooth ride for about two miles the walls were like glass and the floor was covered in dark granules that crunched under the wheels, lighting came from similar lights to the ones on the spaceship but they only lit up on our approach then went off when we had passed, that must be to save power.

I noticed Venetian blinds at regular intervals suspended near the roof piled high with white powder, I puzzled over these and in the end had to ask what they were for, Tapping Worra on the back I motioned her to switch her translator on, she answered my question by saying they were safety measures, seeing me frown she explained, using high frequency cutting creates very fine dust which often explodes, when that happens the shock wave hit's the slats spilling the white powder which cools the flame following the shock wave effectively putting it out.

At first I thought she was pulling my leg, then realised it must be true, it's a brilliant idea, talk about fighting fire with fire, we could use this back on Earth drop big containers of powder with bombs inside it to explode above forest fires, if it worked it would save a lot of grief and forests, next I asked where all the miners were, only one car had passed us going out, with a load of rock, now we had brought food I thought they would all be back at work, she didn't know and before I could ask any more questions we arrived at the end of the tunnel, expecting to see gangs of men toiling with picks and shovels, well this was supposed to be a penal colony, instead all I could see were two men attending a large machine, as we arrived they were adjusting it to take another six inch deep slice of rock off the length of rock face, The machine was just a ten feet long six foot high and five feet wide box with a large caterpillar track wrapped around it with rows of six inch bars protruded from the track, at six inch spacing.

The men having positioned the machine, motioned for us to stand clear and pressed the starter, expecting an ear shattering burst of sound I was pleasantly surprised to hear a low hum, the first lot of bars dug into the rock face after they had passed around the curve of the driving sprocket, the whole machine moved forward travelling along the face, as the bars exited at the back after the curve of the sprocket they cracked off the rock which shattered and dropped on to a conveyor attached to the machine and lifted into one of the three wheeled trailers, I was mesmerised, it all looked so simple, I was interested to see what would happen when the machine reached the end of the rock face, no trouble at all it just reversed and came back cutting all the time, now I could see how the pillars of rock were left supporting the roof.

When it came to a stop at the end of the run, I leaned forward to feel how fast the drilling bars were vibrating, the chap standing near me shouted and knocked my arm away with a metal bar he was holding, this really hurt and I was about to punch him when Worra grabbed my arm, the chap didn't say anything he wrapped a piece of material around the bar to insulate it from his hands, then rested the bar on top of the drill, a high pitched whistle sounded and I looked around to locate the noise, when I looked back the end of the bar was falling to the ground, that was a one inch metal bar and the drill had cut it like butter.

= CHAPTER THIRTY FOUR =

These people certainly know how to use high frequencies. this will be one of the first questions I shall ask Merlin when I get back to Factory looking around the machine now it had stopped, I thought if it was tipped on its face it would make an ideal ploughing machine, we wouldn't need the high frequency stuff, its own weight would do the job, as it went forward the machine would run over the bars pushing them into the ground, then as they came around the rear sprocket they would rip up the ground and help to propel the machine along.

I have visions of a first world war tank with very wide tracks meeting in the middle with twelve inch bars sticking out like a hedgehog digging into the ground running at high speed over the fields churning up the ground, at least it wouldn't get stuck in the mud, better still if the bars were retractable the machine could be used as a normal caterpillar tractor which would save on ground compacting and less cost of tyres.

Before leaving the face I'd noticed another machine covered in dust which looked like a scrapped one thinking at last I had found something worn out I asked what was wrong with it I was told it was kept for reserve, the one unit was capable of keeping the power stations running, if they mined more than they needed it would have to be stored in radiation proof containers.

I'd forgotten about the radiation risk down here, we had been issued with protective overalls and helmets but it doesn't seem to be a problem that worries them, mention radiation back home and everyone panics, we use layers and layers of protection, even luminous watches are banned, what radiation they emit is minuscule, to what I must be absorbing here, the drink I had to combat it must be very powerful, if I can take some back with me it would be worth more than anything else, thinking of valuables I asked Worra if they ever found gold down the mines, yes she said all the power tubes are coated with it we use it for lots of things, we also send some to Earth with our agents, they found it invaluable for barter so they always took plenty with them.

This was another part of my puzzle fitting together, I had wondered how Old Tom their agent on Earth had managed to buy the toiletries I had pinched from the spaceship, now I realised he got his money by selling gold to pay for taxis to travel to meetings with spaceships, next I asked her if she could get me some as I'd never been rich enough to own any, that's easy she said, going over to the chap with the bar she borrowed it, then examining the tunnel wall she found a place and dug a dark looking piece of rock out and brought it to me, it didn't look anything like gold, but when she put it in my hand, it was only the size of a fist but it weighed so heavy, she went on to explain it doesn't look gold because of the artificial lights down here.

Christ I couldn't believe it this lump must be worth thousands and its lying all over the place, Worra had no trouble finding this piece, jokingly I asked her if she could find me a few diamonds to go with it, no she said this disappointed me but she carried on, you will have to get your own, its in big lumps, when struck they shatter and inflict nasty cuts, was she joking, I couldn't be sure, these people had a sense of humour aptly demonstrated when Valda attacked me at the cabin and the Outlaws all fell about on the floor laughing Worra returned the bar and spoke to the chap he disappeared down the rock face and reappeared a few moments later with a large lump and handed it to her she gave it to me.

"There you are" she said "but don't break it up near me."

" no "I replied "I have a better idea, get your man to cut it on his drill," she took it off me and handed it to him, but he returned it speaking to her in her own language, shaking her head she said.

"It can't be done, as soon as it touches the drill it will shatter into thousands of pieces," that put the lid on that idea, then I asked "how big were the pieces it would break into," she said they would be the same size as these on the floor.

"Bloody hell we had been walking and driving over diamonds about the size of wheat grains, it appears that when the drill touches a piece of diamond it shatters and falls on the floor, with the passage of the loaded trailers it tends to get swept to the sides."

My mind couldn't get around the fact that diamonds are treated like ordinary gravel, if they wanted highways here they would use them to pave the roads, I asked Worra how there was so much, she said millions of years ago, this planet hurtled through space getting hotter and hotter, passing through strong gravitational fields the pressures inflicted on it were enormous, it eventually slowed down and was captured by the gravitational field of the Homeworld very much like your world captured the moon in its orbit.

Scooping up handfuls of diamonds I sifted through them sorting out the biggest and filled my overall pockets, Worra wanted to know why I wanted so many, I told her, on Earth we polish them and make beautiful jewellery which cost a lot of money, or in her currency many obligations, she nodded at this and said she had seen pictures of women wearing shining stones in their ears and around their necks, I asked her about the pictures and she said they were in books her partner had given her on her visits to Earth.

I was surprised at this revelation, I hadn't thought to ask her about previous visits, now it appears she had been there lots of times, to collect animals and drop agents, but now trips were too dangerous, agents dropped on one half of my world, were never heard from again, now Worra was talking about Earth I asked when would she be going again, as far as she knew an animal was still needed to replace their old bull, after more questioning I found out bull calves born here are sterile, I expect it's because of the drugs present in the food, if it stops humans breeding it must also be in the food chain which stops animals breeding as well.

Finally emerging into daylight we made our way to the miners rest rooms, there about twenty miners were engrossed in watching something at the far end of the room, threading our way through the crowd, I was amazed to find they were looking at a football match on a big screen Autobrain.

This was incredulous I'm definitely dreaming this time, and to cap it all it was the match we played at the factory a few days ago, pinching myself and Worra just to make sure I found it was all real, How the hell can they do it, we have to have special lighting, production crews, cameras and a whole army of technicians, just for one match, I was that mesmerised I stood and watched the rest of the match to see what was going to follow it, Good God it was only the match played by the Outlaws, no wonder each succeeding instruction match had got easier, they had been watching games on the Autobrains, I cant understand why no one told me, they had seen football matches being played, Worra seemed genuinely surprised and said the system must have been set up on the Founders instructions because it had happened so quickly.

Watching the game my mind was in turmoil, if they can record matches being played without us knowing what else can they do, I remember they couldn't track me in the factory, and lost me when I was inside the spaceship, the Founders had said we wouldn't need cameras to show games on the Autobrains but I hadn't taken any notice at the time, I wondered if Worra and the Duchess knew we were under surveillance and it's the reason any show of affection frightens them,. thinking along these lines it dawned on me, any show of affection was OK as long as it took place in the dark, first in the Outlaws hut with Valda and later with her and Worra in the log cabin, now another thought struck me I still don't know how Merlin knew I'd been bitten by a dog, we hadn't mentioned it, does the Founders have control of Merlin, they soon managed to organise the Autobrains to show football games.

The match on screen ended with the players surrounding me asking to keep the ball, then the screen went blank, the spectators prepared to leave, they had been so engrossed in the game they hadn't seen us arrive, on recognising me a shout went up and I was immediately surrounded with them all trying to speak at once.

I didn't need an interpreter to know what they wanted, within ten minutes we had the men outside and organised into two teams, these men must have watched all our games they settled down straight away, they knew all the rules and regulations much better than I could have taught them, it seems these people have been starved of recreation and were now determined to make up for it, I was getting worried in case it interfered with work but Worra said the supply of ore from the spaceship would make up for any loss, we could empty the ship and get it ready for space travel, Bloody Hell she had mentioned getting another bull, and now she's talking about using our own spaceship to go to Earth to get one.

Watching the game taking place with Worra as referee, didn't interest me any more, my mind was on other things, could I trust hurtling through space in a ship that had spent hundreds of years in a scrap yard, was the ship one of the latest models, and what about the ghosts, I'd been so frightened the last time I was in it, even though the Founders said they were figments of imagination.

If Worra;s story is true I need a plan of action to make sure I get on that ship, first I have to decide whether to take Worra into my confidence and risk her informing on me, or do I bide my time and see what happens, in the end I decided not to say anything, perhaps I could offer my services to kidnap old Bill our prize bull, the uncut diamonds I take back would soon buy another to replace him, I never thought I would feel so exited about going home, and never expected an opportunity like this to come along, perhaps I'm getting exited over nothing, I don't even know if its true, the only way to find out is to return to the factory to see if the spaceship is being fitted out, In my mind I was already planning what to do, and what to take with me, and how Id manage to get left behind when they left Earth.

We stayed the night at the mines and got up early, I wanted to get back to check progress on the spaceship, heading for the rest room to collect our belongings we heard shouting and discovered a group of miners watching a repeat of the games we saw the night before which lasted over an hour, it looked like the programmes were being transmitted every eight hours, we were soon recognised and it wasn't long before we were back on the makeshift pitch playing football again, The Duchess had organised the management into supervising the games which left us free to load up, which suited me the Miners so engrossed in the game didn't bother to see us off and I wanted to make sure my gold was aboard.

Arriving back at the factory I couldn't wait to see how far the preparations had gone in getting the ship ready, dropping off my passengers I drove straight down the wide alleyway between the machines and out through to the back doors, there I found it bustling with activity, the crates of ore from inside the ship were piled up ready to be transported a patch of ground had been cleared which I assumed was for a space ship to land, I soon found Rudo and he explained what they were doing, a ship was due in two days time bringing programmes to up date the auto brains and fuel for the pumps.

Exploring through the ship and watching the men clearing it out and checking the controls gave me a feeling of exhilaration, this was my ship and there was no way they were going to keep me off it, thinking hard i tried to find a way of making sure I'd be included in the crew, I realised my biggest asset was I could control the bull they were going to collect, besides they had no agent to guide them this time the bull may not even be in the area, and I could locate him without raising suspicion, an idea began to form in my mind which seemed impossible at first, but the more I thought about it the more feasible it became.

If my plan fails it means i spend the rest of my life here a virtual prisoner, if it succeeds both worlds are my oyster, I know all decisions are made on the Homeworld and relayed to those in charge here, which happens to be Worra The duchess, and Rudo

Calling them together I took them into my confidence and asked them to put forward a request for me to join the ship on its first flight, I told them to stress the fact I could be useful to locate the bull and control him, also I had saved the ship in the first place, and for good measure I had rescued another ship from the scrap yard.

The three switched off their translators and had a short discussion which seemed to go well in my favour Worra went straight away to the Autobrain linked it to Merlin and sent the message, then came to tell me the answer would come the next day as a meeting of the high council would have to be called, I wondered if the founders would be consulted, surely decisions on crossing the levels would have to be sanctioned by them.

The answer came just after midday Worra appeared quite excited which was very unusual rushing to tell me I had been included in the crew, this came as a surprise I had been anxious all night thinking of reasons why they wouldn't send me, they know I came from Earth and would want to return one day, trying to reason the turn of events it occurred to me this could be their way of sending me back, perhaps they think I may be a trouble maker and will help Worra in their efforts to break away from the Homeworld and to go for independence, it could have worried the authorities, its no good asking Worra or the others if they have orders to leave me there, they do as they are told, their talk of independence could have been a trick to find out my intentions.

It even crossed my mind they were using my obligations to pay for my Earth trip, My Earth trip, that sounded really good to me, although I had the life of Riley here I still missed the comforts of home, I'd been wondering what people were thinking about my disappearance, the regulars at the Virgin Tavern would milk the reporters for all they could, especially if the foxes had left a few bones belonging to the chap the bull had killed and people thought they were mine.

The burns on our bull would take some explaining also the disappearance of my gun and kit could lead to a murder inquiry but after a few days it would all be forgotten with new headlines grabbing the attention.

The number of men working on the ship meant it wouldn't take long to get ready, this turn of events had caught me unawares, I'm not ready by any means, I haven't seen the power station yet and there must be lots of factories to explore on the other side of this planet, although Worra had said there were only these mines and factories, I hadn't even collected anything to take back that's if they let me, tomorrow I must pack as much gold in my rifle barrel as I can I'm sure they will let me take that with me, I only wish I could take a flyer, but I doubt if it would fly with our denser gravity.

I've changed my mind about taking anything back to prove where

I've been, I wouldn't be believed anyway, imagine me telling the authorities I had been teaching the natives here to play football that would certainly put the lid on it and I'd be certified.

Trying to get information out of Rudo or the girls only showed they had no idea when we were going with all the advanced technology here there were no clocks or devices to tell the time i suppose it was the absence of everyday things we take for granted that made me feel homesick, the Chinese water clocks I made to time the football matches, hadn't evoked any interest in the players, Worra asked why I wanted to know when we were leaving, I told her I wanted time to get all my belongings together, she looked puzzled there's no need to take anything she said anything we need will be on board, this could upset my plans for a start I'm taking my diamonds even if I have to swallow them.

= *CHAPTER THIRTY FIVE* =

Getting my lumps of gold aboard would be another matter they were heavy and I had no means of breaking them up Worra had told me agents going to Earth had taken pieces of gold with them for barter, thinking about these agents made me wonder what if I offered my services to take Old Toms place and get information for them they could pay me in gold, that would satisfy all of us.

They would get all the information they wanted and I could get a good living without working hard I may even buy my own farm which I can tell the Council will be a good cover for gathering information, and would provide a safe haven for them to land when visiting.

The more I thought about it the more attractive the idea became, it wasn't as if I was being a traitor, if I didn't do it they would only send someone else, my way I can censor the information I send them, in effect I would be acting as a double agent, except our government wouldn't know what I was doing anyway, and if I was found out, I couldn't be prosecuted, the case would be laughed out of court.

Having made a decision to approach the Home World Council with my proposal and seeing the small amount of work left to do on the ship, I lost no time in finding the Duchess and asking her to act as my negotiator, to my amazement she readily agreed, she thought it was a good idea and could see no reason for the Council to reject it, I started to have my doubts if they turn me down I can say Goodbye to my return home now or in the future.

The Duchess and Worra seemed to take ages to get through to the Council, then we had to wait for them to have a meeting, eventually they told us we were to go before a committee at the Interrogation centre to apply for the job in question, an appointment was made for next day but I had to make an excuse, I said the spaceship was landing next day with spares and supplies for the ship, I told them I was needed to help unload, I was stronger than the people here and could lift heavier loads, this request seemed to impress the Council, they thought I was keen to help, but it wasn't that, I wanted to be here to see the once in a lifetime event of watching a spaceship land.

The next day couldn't came quick enough, I awoke just after sunrise, I say sunrise although the sun wasn't visible through the mist, the day started with a glow in the sky which gradually grew brighter the damp mist cleared half way through the morning and the sun broke through, since I'd been here every day had been the same there doesn't seen to be any seasons or different lengths of the days, so far its been like a paradise and I must admit I enjoy it.

I had asked the girls to warn me when the ship was due as I wanted to see it land, looking back I should have taken notice of the funny look they gave me, Worra had said you will have to be quick but I hadn't taken any notice of her, the workers had got two cars ready by the cleared space and I joined them, looking at the skies I hoped to catch a glimpse of a ship coming in Worra shouted looked at her, then turned to see where she was pointing at.

I jumped back in alarm there before me sat a bloody great spaceship, I hadn't heard anything or seen it appear it just materialised out of thin air, I was shocked, annoyed, and disappointed, all at the same time It made me mad, being the only Earth-man since the beginning of time to get the chance of seeing a spaceship land and I'd missed it.

How the Hell can something as big as this appear in a fraction of a second, the sound of laughter cut short my consternation Worra the Duchess and some of the men were laughing at my expressions, they soon stopped when I said if they were so clever, how was it possible for such a large object to travel at thousands of miles an hour and come to a stop almost instantly, the Duchess answered the ship doesn't actually travel, when Electronic fluid is pumped around the shell of the ship it is irradiated by atomic frequencies just below disintegration level this is modified by the bulk of the ship this separates the atoms from their shadows which can be reformed on different levels, then they are programmed to recombine at coordinates planned by Autobrains.

I don't know which surprised me most, the fact she had volunteered to tell me and not Worra or that she knew such technical information, now she was in a talkative mood it seemed too good an opportunity to miss, laughing at her outburst I said don't be silly your making it up you wouldn't know how it works, this annoyed her and she carried on, we all had to study engineering in our younger days and taught about transference.

Seeing my look of non belief, she explained that the science of every atom having a shadow, or two atoms occupying the same space, at the same time, I didn't believe any of this and it showed on my face this made her more agitated and she tried to explain more simply, Everything has an equal and opposite force, hot and cold, black and white, atoms have their positive and negative, these forces interact with one another and keep them in equilibrium our force field separates them, with the result we can direct the shadow into another dimension, then recombine them on coordinates anywhere we want.

She may be telling the truth but there is no way I can be sure, then I thought of a way to catch her out, during the war men operating radar stations had been dying, because the high frequency radar beams were literally cooking their insides, so I asked her why didn't the high frequencies involved in their ships kill the crew She answered that its common knowledge electrostatic energy flies to the outside of a container and in the case of a ship that's where its used up producing the shadow atoms.

Her revelations sounded too incredible to believe, but just because I don't understand them it doesn't mean our scientists on Earth wouldn't, with this in mind i started to take more interest in her explanations and even inserted a few questions, I wanted to know how the ship landed on the ground and not under it, or hundreds of feet up in the air, the Autobrains had only to be a fraction out in its calculations and it would end in disaster, Oh no she said that would be impossible the position is confirmed many times before the other half arrives, I interrupted her if the ships appearance is instantaneous as you say there wouldn't be time to check, Of course she said I forgot to mention time is halted when atoms are apart, they cant decay so time stands still.

If only half of what she is telling me is true its quite a revelation and it got me thinking is she pulling my leg or is this some sort of test to see if I can be trusted to act as their agent on Earth, deciding to take the bait I told her.

"On Earth what you have told me would be of no use to us we haven't the means to make the porous shell or the metal you use, and as for the high frequency radiation we have no chance of doing anything like that for hundreds of years, nor do we have anything like your Autobrains even if I took one back with me we haven't the power to drive it, and there is no way we can get into space in my lifetime, these advances follow a strict pattern of evolution, we normally have wars to advance our science, but now we have the atomic bomb each nation will be frightened to use it for fear of contaminating the whole Earth with radiation.

A look of relief flashed across her face, seeing this I realised she had been checking me out, and I suppose the answers I had given had satisfied her, now she can report back her findings and we can get ready to set off, although I'm sure they will want to train me in the use of their communication devices, unless my translator can do the job or will they take it off me it only works if another person has one, I asked Worra if they were ever taken off she said.

"No she had seen them on live people going into the crushers," this appalled me and I asked why she had witnessed such a horrible thing, she said some people from a far world had taken a starship and had been caught, her school had been lined up as witnesses to a trial and their execution, thinking about this terrible punishment it didn't register for a few minutes, then it struck me she had said people from a far world, this really got my interest and I blurted out. "

"You mean to say there are more human races scattered around the universe, they both laughed "of course there are do you think your Earth is the only one.

Our ancestors seeded many worlds on their voyages through space and their offspring have colonised more as they acquired space travel, your world will soon use the moon as a base to explore space, we expect you to use up your resources getting into space but it will stop you destroying yourselves and unbalancing the universe.

Hang on a minute I interrupted your people didn't discover the secrets of space flight they stole it off another civilization, "No" the Duchess shouted in indignation.

"We found it on this planet, it had been left behind by our ancient ancestors, it may have been by mistake or it may have been on purpose for us to find, we never found the reason why the factories were left here, the discoveries we did find advanced our evolution by hundreds of years If our scientists hadn't been killed in the factory explosion we would now be on a higher plane of evolution instead we are stuck here with no prospects of advancing, the founder can't help us its against the laws of the Universe."

She suddenly changed her tone and speaking more or less to herself she said and yet they came back to see you, perhaps they are helping us in their own way, you have started us to think for ourselves again, the people on the Homeworld are waking up and taking interest now the harmful drugs have been taken from the food chain and the football game has given them an interest, then she turned to me and casually said when you come back will you bring other games that you Earth people play, this completely threw me and I asked her what she meant.

You have to come back she said, you are obliged to Worra, her partner has been dead over ten revolutions, of the Homeworld and she has chosen you for her consort, Christ this is getting bizarre, I laughed "don't I get any say in it and what does it mean," now she surprised me again by saying "since the drugs have gone males have been taking advantage of the females and doing nasty things, I'd heard that expression before, when Worra described the Outlaws to me in the cabin she had said.

"They did nasty things to women if they caught them," when I pressed for more information she refused to speak and went quiet, since then I'd pretty well forgotten about it.

Now was my chance to clear things up I asked the Duchess straight out "what were the nasty things you talk about," she went very quiet wet her lips looked around to make sure no one was listening then leaned towards me and whispered "the men try to revive the ancient rites of creating offspring," this caused me to burst out laughing, these people thought nothing of parading around in the nude, and yet they were very prudish about sex, the Duchess was very indignant and almost shouted.

"Long ago we were too many for our resources and laws were made to keep our numbers down, but they failed and other ways were found, now only council members and technicians are allowed offspring," anyway she added "it won't effect you, Worra is a long way beyond the age limit," I'm sure I caught a hint of jealousy in her voice, but she must be joking about Worra's age I'd seen her in the showers and she had the face and figure of a teenager, it had never occurred to me to ask her age, its just not done to go around asking women how old they are.

Most of the people here look aged between twenty and forty I'd put this down to it being a sort of national service camp where the inmates look after themselves and exchange mined ore for supplies, then after awhile they return home, the Outlaws take their old people to the food farm where they go into the crushers for fertilizer they are then replaced by people from the Home world who have done something wrong and are banished to this colony as a punishment, Worra had said in the old days they were hunted down and killed, perhaps this was before the drugs were introduced and it was done to keep their numbers down.

Leaving the Duchess I went in search of Worra to ask about my engagement and how it will effect me, it appears I've been drafted into Old Toms place, this could be a good deal for me Tom never did any hard work they said he had a weak heart, now I know it was a cover for him not being able to lift anything heavy because of the stronger gravity, and he had plenty of money, he always gave the taxi drivers a good tip when they took him to catch the trains, some of these trips must have been when he made contact with other agents and changed his gold, he used to talk to me about his time in Australia how they rounded up cattle and often had to drag them out of waterholes with cars, now I know he was talking about his life on the Homeworld and the waterholes were the dew-ponds which are scattered around this place.

I think I knew him better than most people but never knew where he came from, we assumed he was one of the hundreds of wartime refugees wandering around the country looking for work, he appeared at the farm just after the war to ask if he could park his caravan in our Rick yard when his car broke down, he slept in his van while he worked on his car but he never managed to repair it, he often did odd jobs and ended up looking after the chickens.

Failing to find Worra I decided to ask the Duchess some more questions, I found her busy sending messages on the Autobrain, she pointed to a message plate lying under the screen, which was evidently meant for me as it was in English, it was from the high Council summoning me to report for our postponed meeting at midday in the interrogation centre, this pleased me it showed our trip was imminent, as there was no reference to the girls I assumed I would be going on my own, The Duchess told me Worra had also received a message from the Council which had pleased her so much she had gone to factory one to answer it on Merlin, it's a pity I hadn't ask what the message contained it would have saved me a lot of trouble later on.

I awoke early to find the usual morning mist swirling around even though I was keen to get airborne I remembered the last time I tried it and resisted the temptation, its easy taking off but once above the mist its another world of cotton wool with an invisible ground to land on, waiting for it to clear I finally took off, I'd left instructions for the girls to drive over in case they were wanted and to bring back supplies.

Flying above the mist was exhilarating like flying over a land covered in snow, glistening in the rising sun as the mist cleared, tops of trees started to appear then the buildings of the centre showed up, flying over the roof to look around I saw the large water tanks which must be for the showers there was also a large pond which must have been created by the overflow off the roof landing behind the centre I intended to have a look around.

I was surprised no one came to investigate, if someone landed in my backyard I would want to know who it was, walking around to the front entrance I entered and began to explore the place, first the shower rooms, there were ten of them which was more than ample for the number of people on this planet, I began to wonder if there were a lot more people living on the other side, or were there a lot more here when the factories were built, this would account for the size of the tanks on the roof, looking around the place I didn't find anything to interest me, I couldn't find any way of getting on to the roof either, one of the rooms I looked into was the dormitory with ten men sleeping there.

So I went back and woke them up, this was a big mistake, trying to ask them directions was like stirring up a bees nest, leaving them all shouting at each other, I don't know if it was because I frightened them, or whether they thought I was an advance guard for the Homeworld Councillors, I escaped and went for a walk around the outside of the buildings.

Half way along the left side I found access to the roof, a series of holes had been cut with a flap covering it on the inside so when climbing you put your foot in the hole pushing it open, now I know why the buildings are sealed its to keep air inside, so there couldn't have been a lot of air outside, that's why the spaceship had all those suits and air cylinders on board.

Climbing up was easy the loss of air when pushing in the flaps with your toes wouldn't have been much, reaching the top and looking over I saw the reason for no inside access, the walls of the building were extended another three feet above the ceilings and the water filled it about three inches from the top, half the tank was covered over with black metal and the other half was shiny silver as if it had been chrome plated, a walk way extended right across to the other side with no handrails i suppose being only three feet deep it was safe enough.

I walked across the walkway and looked over the side there was a pipe going down to the pond which I'd seen from the air, this puzzled me, there was no way the mist could fill a tank this size, what I think is an overflow pipe must me the inlet, working it out I reckon the black half of the tank heats up in the sun, the hot air expands, and exits through a one way valve then the cold night air contracts it, creating a vacuum which sucks up water, when suction drops a valve shuts off the water, and lets in air to start the cycle again, the pond must be connected to other dew ponds through the water table, otherwise if it dried up it would break the suction.

= *CHAPTER THIRTY SIX* =

Wondering if there were any fish in it, I lost no time in getting down to check it out, only to be disappointed, there wasn't a living thing in it, the water was completely sterile no wonder it was crystal clear, all my hopes of a few days fishing were dashed, never mind, I consoled myself with the thought Id soon have all the time in the world, or rather my world to fish as much as I liked.

A loud whistle brought me back to reality, the Duchess had arrived and was demonstrating she had finally mastered the art of whistling, she told me Worra had been summoned to the meeting, but she hadn't arrived yet, then she dropped a bombshell Worra will be staying on Earth with you that must be the reason I'm obliged to her it must be a form of marriage, part of their moral code, if I want to go home I've no option than to take her with me, this latest twist is going to cause me problems its bad enough having to explain my disappearance without the added burden of a wife to complicate matters.

An attendant spoke to the Duchess "come on she said.

"We mustn't keep them waiting," following her I found the council members already seated, I recognised all five of them straight away they were some of the chaps we had rounded up to save from the dogs, when the leader waved for us to be seated I knew this meeting was going to be a lot different from our previous ones, I got the impression they were obliged to me and it showed in their attitude, even the Duchess who was acting as interpreter appeared calm and collected, much different to her anxious attitude when in the presence of High Council members.

Their spokesman started off by saying how pleased they were to have me as their agent on Earth then he went on about me saving two valuable space ships and entered two factories, he even brought in the fact I had introduced the game of football, while he was singing my praises I began to wonder why all the fuss, did he know something I didn't, I was beginning to feel like a Kamikaze pilot about to go on his last mission, it only wanted him to walk around the table and tie a handkerchief around my neck to complete the scene.

They asked if I had any questions, I said Id have difficulty explaining Worra to my friends, he answered saying she resembles the races on the far side of your world, tell them she came from there, he went on to explain, her task is to contact our missing agents, and you will assist her in this by escorting her wherever she wants to go, now I was alarmed do they expect me to go into China or Russia.

"Hang on a minute ", I said," we cant go anywhere we like on Earth, we need passports and permits," we know," they said "that's why you were picked you showed the Founders you have the qualities to think for yourself."

Bloody Hell if the founders are behind this charade why didn't they come out into the open and ask me face to face instead of going through all this play acting, next I asked when were we going to set off, he tried to evade my question by changing the subject and I had to press him to give me an answer, eventually he admitted they were having trouble programming the Autobrain, all the star charts in the memory were four hundred years out of date.

I could understand this it had taken us days to update Merlin when we switched him on the first time, then it had taken three days to start up the machines in the factory, I was interested in the fact that the star charts were four hundred years old, no wonder the bodies on the ship were bloated if they had been entombed in the ship for that length of time, I still smile at the antics of the Outlaws trying to get the clothes off the blown up corpses, I told Worra to get the men to puncture them, but I'm afraid I didn't have the stomach to stand around and find out what happened.

Now I know I have three or four days to get myself ready I feel more at ease, there's so much I want to do I feel I've completely wasted my time here, anyone in my place would have asked Merlin about a cure for cancer, or how can we overcome radiation poisoning, or better still how can we create fusion of Hydrogen atoms, instead I've messed about with the Beale treasure Code and other silly things, I remember the answers Merlin gave me concerning God, and how the two girls laughed when I told them about our beliefs and how our world was created.

"Do you want to ask the Council any questions" the shrill voice of the Duchess cut into my day dreaming, " Yes "I shouted before I had even thought of anything"What does it mean when you say I am obliged to Worra", the councillors went very quiet and looked at one another as if seeking support for their answer, the spokesman seeing he was expected to reply cleared his throat and said "when a man has partnered a woman for ten orbits of our planet he has the right to own her if she is willing, this is a rare privilege and the women don't take to it lightly, Worra must have much respect for you as she must submit to your every wish", "Wait a minute "I interrupted," if that's the case how come she's In charge of me on Earth."

That's easy" he replied, "On Earth you revert to your own laws, on your world women are the dominant specie, and she will be acting on our orders.", I didn't see any point in arguing so I let the matter drop, the spokesman had one last bit of information for me he said I will have to keep my translator on to communicate with Worra until she learns my language, if anyone tries to remove it tell them it is a religious symbol, how naïve can these people be, imagine me being interrogated by the KGB, and telling them I cant take it off on religious grounds, I don't think my head would stay on my shoulders for very long.

The spokesman suddenly stood up clapped his hands and the committee all walked out, I had expected them to wish me luck or something, but to see them just ignore me after the glowing references they gave didn't seem right, although by now I should have known what to expect. With the meeting over all I wanted to do was to get my kit together with priority for the diamonds, the Councillors hadn't mentioned what I could take with me, I hadn't asked in case they said I couldn't take anything, what they don't know they wont grieve about.

The Duchess was waiting expectantly for me to accompany her but I was more interested in finding out why Worra hadn't appeared, its a death sentence disobeying an order from the High Council, which made me concerned about her safety, the Duchess had contacted number two factory who confirmed Worra had left just after her.

It would be dark soon and if she has crashed it wouldn't take long for the dogs to find her, and I wouldn't like that to happen to my worst enemy, leaving the Duchess to organise search parties from each end I set off to search for her.

Flying into the glare of the setting sun didn't help, in the end I flew in overlapping circles this enabled me to see better but it meant covering less ground, straining my eyes because of the approaching darkness and flying quite low I soon saw her crashed flier, landing beside it I saw no sign of a body this cheered me a little it showed she had managed to walk away from the crash, which was over halfway from the factory, now I had to decide which way she had taken, being aware of the death sentence hanging over her for non compliance of orders she would have made for the interrogation centre, setting off in that direction shouting Worras name, I hadn't gone far when I was relieved to hear an answering voice, I wasn't so relieved to hear another sound the sinister rustling in the undergrowth my shouting had attracted some unwelcome visitors and they were only waiting for enough of their pack to appear before rushing me, turning I waited for a dark shape to show itself then fired the flash lit up his eyes as he reared up and fell over backwards into the path of the pack who had been silently creeping forward to make a concentrated rush at me.

The dogs reacted in their usual way attacking the unconscious body with a ferocity that sent shivers down my spine, Christ if I'd waited another second I'd be the one under that pile of fighting dogs, they were so mad with their blood lust snapping and snarling they totally ignored me, this gave me precious minutes to locate Worra, she had climbed the biggest tree in the area which wouldn't have afforded much protection I called her to follow me and she jumped down beside me without any hesitation this was a relief it showed she wasn't injured.

Grabbing her hand we ran around the tangle of fighting, dogs I was almost bowled over by a latecomer rushing to join the fray, I shot him from behind and his impetus carried him right into the core of the fighting, this was like pouring petrol on a fire, the fighting roared to a higher ferocity, this distraction was a god sent opportunity, steering Worra to my flier I had left it ticking over and shouted for her to board while I delayed the dogs she hesitated and I had to shove her in the back, her wrecked flier lay upside down nearby, ripping off my coat I crawled under it and unclipped the fuel cap.

= CHAPTER THIRTY SEVEN =

The fuel gushed out soaking the coat I'd thrown under it, in the background I heard the flier speed up the dogs had also heard it, the noise of fighting died away to be replaced by that horrible baying sound dogs make to terrify their quarry, they were coming after me, stepping back I took careful aim and fired.

WHOOSH, Bloody Hell I hadn't expected it to explode so violently, The sudden flash of light blinded me, I could hear the flier but I couldn't see it, calling to Worra that I couldn't see and for her to shout me directions, she did better than that she jumped down grabbed my hand and rushed me into the pilots seat my head had cleared a bit and I could hear the dogs howling in terror in the distance as they fled from the fire.

Telling her to take it easy now the danger was over, we didn't want to spoil it by crashing again, sitting on my lap she managed to take off without mishap rising above the trees I asked if she could see the lights of the factory to guide us, after a while my eyes started to clear and I could see the stars, this was a relief I thought I had damaged my eyes, by the time we reached the factory my eyesight had returned to normal.

Rudo was waiting to meet us when we landed and everyone wanted to know what had happened, I understood what Worra was saying as she had left her translator on, she gave a dramatic story of how I had saved her from the dogs without any thought of obligations, I felt quite proud of myself listening to her, then I remembered the part she played, she had left the safety of the flier to help me.

Wait a minute I said forgetting the crowd couldn't understand me, Worra was very brave she saved my life at the risk of losing her own, Rudo translated this to the crowd, well I know these people had changed out of all proportion since I came here, They actually clapped her, this was the first time I had seen them show so much emotion that I asked Rudo where they had learnt it from, he said they got it from watching old recorded football games agents had brought from Earth many years ago, it appears they were now searching through old archives to show interesting items on the Autobrains.

Rising late and rushing breakfast so I could feed my questions to Merlin, I found I was too late Worra was using the Autobrain, instead of waiting around I decided to fly to the cabin and collect my rifle and a few other belongings, surely nothing could go wrong on this short trip, taking extra precautions and making sure I had spare fuel a couple of lamps in case I met the outlaws, and spare Tee gun, I set off on my own.

I still get a sense of pleasure flying above the mist I'm in a world of my own only this time it was more pronounced I felt this was the last time I would ever fly one of the fliers, guessing the position of the cabin I prepared to descend through the mist, previously I'd found that by going down very slowly the down wash from the rotors pushed the mist away, this sounds easy but it took me five attempts before I found the cabin.

Landing beside it I unloaded enough supplies to make myself a meal, and a drink then I intended to collect my things and take off, long before darkness set in, so far so good nothing can go wrong stoking up the fire and getting my meal ready in these warm comfortable surroundings started me thinking of the first time I came here how much things have changed, I also remembered every time I daydream something drastic happens, and this time was no exception.

Collecting my rifle and a few more odds and ends I tore strips off my old shirt and used it to tie my gun to the flier, engrossed in tying it on securely I became aware I was being watched swinging around I felt so foolish there was no one there, laughing at myself for being so silly and putting it down to the absence of noise I returned to the cabin pouring a drink I sat looking around in case I had missed anything, Of course there was, Old Rip Van Winkle had left his name etched on the beam above the fireplace, it was only fair I put my name up as well, using the red hot poker out of the fire I wrote my name and the date I came here, leaving the fire to burn itself out I made for the door, reaching for the latch I recoiled in horror.

A large bloodshot eye was looking at me through the latch hole, Bloody Hell I nearly fainted with shock, I swallowed hard, waited till my heart slowed down then called out, after the third call an answer came back Valda, Valda, Valda, this sounded suspicious to me, if it was the Outlaws we knew why hadn't they just knocked on the door, standing in the doorway thinking it over the only way to find out what they wanted was to confront them, flinging open the door I was surprised to find a dozen Outlaws standing there one. stepped forward said Valda and motioned for me to follow him.

There's no way I'm going to their village, it would take hours to get there, I'm leaving for Earth in two days time, and I wasn't going to let anything stop me, I tried to speak in their language but it was hopeless then they started getting nasty pulling at my clothes trying to drag me with them, this situation was getting out of hand I'd never seen these people get so worked up before, they were determined to make me go with them and I was just as determined not to, pushing my was through the crowd I made towards the flier, now they turned on me, in desperation I pulled out my Tee gun and waved it in the air, talk about waving a red rag to a bull it had a totally opposite effect to what I expected, the whole crowd attacked jumping on me so fast I collapsed under the sheer weight of bodies, terrified I lashed out in all directions, feeling my fists and feet making contact I managed to get back on my feet just in time to see a red beam hit one of the chaps near me, he squealed and fell unconscious, I glanced around to see who had come to my rescue only to be hit by the same beam and knocked out, I spent the next hour being carried slung under a pole, to their village, waiting for them to untie me I made a grab for the Tee gun, but my arms were stiff from being tied I fumbled it, and got shot again.

Waking up with a terrible headache and a thirst to match I tried to get up off the bed I was lying on only to fall over, lying on the floor I waited for my head to clear, this seemed to take ages, even in my befuddled state it felt wrong, I had been hit before and it hadn't felt like this,, the pain had been the same as I remember, but the after effects had been very different.

Eventually I managed to stagger to my feet and looking around, found I was in one of the Outlaws round houses, there was even one of our oil lamps burning to light up the place, because it wasn't very bright I failed to see another occupant lying in the next bed until I nearly fell over it, picking up the lamp and shining it on the sleeper I was surprised to find Valda looking up at me, she tried to speak but I could see she was in a bad way, she was gasping for breath and her skin looked deathly pale, seeing her struggling to breathe I realised I was having the same trouble it felt like I was in a submarine with no air, Bloody Hell that's it I'm being suffocated the lamps are using up all the oxygen I tried to force the door but its been blocked solid.

Banging on it had no effect I don't know if its day or night outside, realising I hadn't much time before the air ran out completely brought me out of my stupor and my brain raced looking for an answer, the roof, I'd been told these houses had been built by visitors who had lifted the roof on, dragging my bed to one side I stood on it and pushed with all the strength I could muster the bed promptly collapsed, under my weight and the extra pressure.

But it had lifted about six inches and I'd felt a gust of fresh air, and seen it was dark outside, using my bed like a pit prop I shoved the roof up and wedged it, having no time to be soft I tipped Valda on the floor and used her bed to prop up the other side, pushing the bed sideways tipped the roof over and it landed with a crash outside.

Pausing to take a few lungful of fresh air I righted Valda's bed and put her on it, not being sure what to do with patients suffering with carbon dioxide poisoning I turned her on her front and gave artificial respiration, she gasped a few times then I heard her breathing change for the better she even tried to talk, but I couldn't understand what she was trying to say.

I put my finger to her lips for her to save her breath, I took deep breaths and mimed for her to do the same, expecting her to have a worse headache than mine I got her to sit and I massaged the back of her neck, this worked wonders and in no time at all she was walking round the room.

Then she pointed to the V shaped openings and made a yapping noise like a dog I knew what she meant we were in danger if the dogs came around, catching hold of my hand she picked up the lamp and led me over the side wall to another hut about fifty feet away, this one was empty I suppose the normal occupant was away on a hunting trip, this was confirmed by the fact there was only one bed and clothes and weapons were missing.

Valda closed the door but I wedged it open to let air in, I didn't want a repeat of air poisoning, she nodded approval, I only hope there hadn't been any fatalities because of our lamps, No wonder the Outlaws had been so keen to get me here, I don't suppose they had any idea what was causing them to be ill especially when they don't seem to suffer from any diseases of any sort.

Noticing Valda was shivering in the cold night air I motioned her to take the bed, I turned the lamp down and prepared to spend the night on the floor, she leaned over and literally pulled me into her bed, I panicked I didn't know what to do, after my experiences with Worra and the Duchess this action was quite alien to these people, Then I calmed down she was only doing it because she was cold, Well that's what I kidded myself to think, her next actions dispelled any doubts in my mind, wrapping her arms around me she started to nuzzle my neck, I pushed her away but it made no difference she was determined to have her wicked way with me.

I prepared to jump out of bed, I couldn't take advantage of a young woman who was at deaths door a short time ago, then I had second thoughts she was so insistent her antics were having the desired effect and what followed next was only to be expected in the circumstances, later I had pangs of conscience but I soon dismissed them by saying it was in the cause of occupational therapy the exercise certainly improved her breathing and got rid of some carbon dioxide.

Well that's what i excused myself with and I'm sticking to it, dropping off into a contented sleep I passed the night in sweet oblivion, only to be rudely awakened by loud voices outside untangling myself from Valda's tight embrace I peered through the gap in the door, a dozen or more villagers were gathered around our former prison, no doubt wondering what mysterious force had lifted the roof off and spirited us away, not knowing why they had locked us up in the first place, I was loathe to call out in case they locked us up again, thinking it would be better if Valda did the explaining I shook her awake.

Big mistake, she grabbed hold of me to drag me back into bed, I only knew one way to break her hold, leaning over I kissed her lightly on the lips, she leapt up in alarm instantly awake, I don't know why these people are so frightened of being kissed, perhaps it's a fear of germs or something, putting my finger to her lips to silence her, I led her to the door she saw what was going on, and called out to them, the crowd surged forward shouting excitedly, I was fully prepared to go down fighting for my life, but she raised my arm and I realised they were honouring me.

Valda took charge of the situation getting the roof repaired and organising breakfast, this was really living the life of Riley, till I suddenly remembered I was supposed to be returning to Earth tomorrow, How the Hell am I going to explain all this to Worra and the Founders, they must be wondering what's happened to me, and what if Valda says anything about our night together, I'm obliged to Worra and if that's anything like being married I'm in big trouble, not only that, how am I going to tell Valda I'm leaving her.

The flier must still be at the cabin but how to get there and get airborne is going to be difficult, Valda must have fallen for me in a big way she hasn't left my side all morning, thinking over the problem gave me a headache and I sat down on a log, idly picking up a stick I drew a picture of a flier, in the sand, this gave me an idea, why not take her with me, drawing two people on the flier I showed her the picture, she had been on a flier and knew straight away what I meant.

Only stopping at the hut to pick up our weapons and a pot of water we set off, so far so good all I had to do now was to trick her into letting me get airborne, I intended to leave her in the cabin where she would be safe, I hadn't realised how far the village was from the cabin and it was midday when we arrived, my flier was still parked outside, and there was another one parked beside it.

Hailing the cabin I was relieved to hear Worra answer she came running out to greet me like a long lost brother, then ignoring me she turned to Valda and started to talk excitedly in her own language I could only understand parts of it and had to guess what they were saying, guessing right I was prepared when Worra swung around and snapped at me "What does she mean "you are obliged to her", "Me obliged to her", I said falling back on the old trick of denying everything just to give me time to think of an excuse, I continued "she's grateful to me for saving her life, she was dying from the poisonous fumes of the lamps we gave them, and I felt it was the least I could do to help, you know its in my nature to help others without thought of reward or obligations, its different for us on Earth so you can't blame me for anything," feeling very smug and satisfied with my clever handling of a potentially dangerous situation I turned to enter the cabin.

Crash, a great weight landed on my back propelling me through the doorway and flattened me to the floor, it had happened so fast I only just managed to turn my head before I hit the ground otherwise my nose would have been squashed, my hair was pulled so hard I thought my neck would break, then an arm came round and started to throttle me, no matter what I tried the weight on my back wouldn't budge, rolling over in desperation I was on the point of blacking out when a loud scream drilled into my brain and the strangle hold relaxed, the weight rolled off my back and I lay gasping for breath, while a storm of shouting and screaming erupted around me.

Staggering to my feet I watched in disbelief as the two lithe figures of Worra and Valda fought like cat and dog, scratching biting and ripping lumps of hair out of each other, I was mesmerised these normally placid people certainly knew how to let their hair down, or rather rip it out, standing there getting my breath back I couldn't make out who was who, waiting for a break in the fighting or to be more truthful to let them tire themselves out, I waded in and grabbed one of them.

Luckily it was Worra she was the one who had attacked me, bending her over my knee I whacked her behind so hard it hurt my hand, she screamed out in pain and anger, another figure came to help me, but I grabbed Valda put her over my knee and gave her the same treatment, just this was to show there was no favouritism, looking at the two dishevelled girls on the ground covered in dust with little rivulets running down their faces where their tears had washed away the dust I burst out laughing, they looked at one another and realising what a sight they presented and seeing the funny side joined in.

I ended up with a nose bleed and the girls ended up the worse for wear, but my actions had certainly saved the day, by the time we had cleaned ourselves up dressed our wounds and had a meal it was getting late, not risking a crash in the dark we decided to stay the night in the cabin, I was a little apprehensive about sleeping arrangements, had Valda told Worra about our arrangements last night, evidently she had, because they launched a plot to have their wicked way with me, Still being too embarrassed to undress in front of the girls I went into the back room to change, I heard dragging noises come from the other room but ignored them, entering the dark bedroom my arms were suddenly grabbed and I was flung onto two beds that had been put together. I panicked and tried to fight them off, I thought they were going to beat me up in retaliation for spanking them.

However they had something entirely different in mind, one held me while the other ripped all my night clothes off, by this time we had become so entangled I didn't know who was who, and I didn't care, it crossed my mind that something had turned them on, the two lithe naked bodies pressing against me certainly had an effect, and I made the most of it, whatever had changed these girls from frigid females into such wanton women, must be very powerful, if slapping their bottoms was the reason, they are in for a spanking good time in future, the rest of the night, passed in a blur of every man's dream two virile young women in bed without a shred of embarrassment trying to out do each other in pleasing me, I don't know who won the contest, but I do know I enjoyed myself beyond my wildest dreams.

= CHAPTER THIRTY EIGHT =

Waking up very late tired from the night before, I found the girls chatting away as if nothing had happened between them, although by their appearance it showed they had been in a fight, both had big scratches on their faces and lumps of hair missing, I'm surprised they were even talking to each other.

Perhaps Valda hasn't told what really happened between us, but then there had been no objection when we shared the same bed last night, and at times in the dark I wasn't sure who was who, anyway its no use worrying, Valda's staying behind in the cabin until the next hunting party comes along and I'm off back to Earth, I've been enjoying myself so much I had completely forgotten about the spaceship I even toyed with the idea of not getting back in time, the snag is I'd never get another chance.

Now my minds made up I couldn't wait to get back to the factory, loading up the fliers and helping Worra to get airborne didn't take many minutes, Valda was standing by my flier and I naturally put my arm around her and kissed her goodbye, only to be slapped hard across the face, it had happened so suddenly I wasn't prepared and it really hurt, I was so shocked I tried to ask her why, but we couldn't understand each other.

It wasn't until I was airborne that I realised she had wanted me to retaliate and that would have aroused her, I suppose with Worra flying off into the distance she thought she would have me all to herself, that was the secret of these people, their emotions had been suppressed for so long their passions were only awakened by violence, come to think its no different on Earth, violence towards women usually end with the woman pleading with the judge to let their attacker off, even if the man had tried to murder them.

Flying around the cabin twice to give Valda the impression I didn't like leaving her, I set off to follow Worra, flying high over the factory I was relieved to find the spaceship still there, the Founders had said it was leaving today, but they never give any time, landing beside her flier I saw the workers had come out to see our arrival, I suppose they were curious to know where I'd been, after we had landed Worra spoke to them.

She must have told them some believable story about her appearance which seemed to satisfy the crowd and they gradually drifted away, Worra now gave me the glad news we were needed aboard at dark, at my reckoning this was in four hours time this planet only has three time divisions light, midday, and dark, perhaps the translators can only interpret sun up and sundown as light and dark, we couldn't survive on Earth without our time measures, here they have no use for them, I wonder how they will progress now I've introduced water clocks for measuring half time and full time at football matches.

Any ship setting off on a long voyage would have been a hive of activity loading provisions, fuel, luggage and personnel, there was only five crew on board including me and Worra, I'd managed to collect my belongings together and hidden my stones and gold.

Although I need not have bothered no one, was interested in what I had, when I told Worra I'd carry her luggage on board she said everything she needed was on the ship, when I finally got aboard and asked where my cabin was Worra translated for me and everybody laughed, she explained there wasn't time to sleep, jumping to another level was almost instantaneous, it was only the transition period when we were getting airborne, landing, and waiting for beacon signals that took any time.

Sitting in front of the big navigation screen I pretended to be at a loss to understand how we could travel so quickly, the Duchess had already explained it but I wanted to see if Worra told me the same story, she told me first I would hear the pumps whine, this was to pump fluid around the shell of the ship, this produced the radiation field which when irradiated to atomic frequency shook the atoms apart this produced the shadow which was projected to a point worked out by the Auto brain and was recombined at that place, they also worked out the next point and at each point the inhabitants of the nearest planets would be able to see us, as our atoms recombined into whole atoms and in this state time reverted to normal.

Time for a journey only consisted of the time spent in the visible, when atoms are apart they can't decay so time stands still, of course I had to say why don't you do the journey in one jump, she said this was too dangerous, space was full of dust clouds and magnetic fields which could cause malfunctions.

A whirring noise cut short our conversation. I waited for the lurch to show we had taken off, when this didn't occur and the red dot moved along the dotted line on the screen I asked Worra why it hadn't happened, she looked puzzled for a moment then said.

"Our lift off from Earth had been programmed to include the weight of the bull with the consequence the gravitation pull hadn't been compensated for." "Wait a minute" I interrupted, Merlin told me we cant alter gravity.

Of course we cant but the Auto brains can, they can calibrate the speed of the electrons orbiting the nucleus, too high and it emits energy, too low and it absorbs energy by keeping them just below the point of emission it vibrates at its own frequency cancelling out the braking effect of the gravitational pull of the nearest mass.

"Wait a minute I interrupted again, can you explain in plain simple English what the gravitational braking effect is.

"Of course," she said," its one of the first things we learn."

"You do know what an atom is," She threw this question at me and I answered "yes" before I realised the importance it meant, she nodded and carried on, imagine a standing atom of hydrogen with its electron moving away from a larger mass gravity field which tries to pull it back, the braking effect on the electron pulls the whole atom towards the larger mass, the electron carries on orbiting its proton and is now travelling towards the gravity field so it has no braking effect, In fact it gains speed making its orbits elliptical and the power to keep going, on denser atoms the effect is more pronounced that's why they weigh more.

As she carried on talking I noticed a few words of her own language started to come through my translator, seeing my puzzled look she asked what was wrong, I told her it wasn't working properly, she explained it was because there were no comparative English words to translate into.

Changing the subject I asked why I didn't feel any different now half my body mass was somewhere millions of miles in space, she laughed and said we don't alter because we are inside the spaceship shell any electric forces fly out from the source, then she noticed the red spot was blinking, "come on" she said, "your home," I couldn't believe her its only been a few minutes since we boarded.

While we were talking one of the crew came over pointed to the screen and handed me a small shoulder bag which was so heavy I dropped it, looking inside it was full of gold nuggets the size of peas, this made me think, if Old Tom had this amount of gold where did he change it into cash, my day dreaming was cut short by a shove in the back, the crewman pointed to the screen again.

The red spot had stopped Worra jumped up and made for the door, I just had time to grab my gun and the two bags before the door opened and I was pushed out, I had to jump to save myself from falling over, and landed with a bone jarring thump, what a difference the change in gravity made, I felt I was carrying a big weight besides the bags I was holding, this was going to take a lot of getting use to, taking a deep breath I found the air was smoky, it had been raining so it should have smelt clean and fresh.

Looking around I saw it was going to be difficult to find the bull, dark clouds were scudding across the sky so there was no moon light to see by, and we had no torches to light the way, this was a stupid turn up for the books, they had all the advanced technology for space travel and Autobrains but they had no portable lights.

The crew had dropped the ramp and were waiting expectantly for us to bring the bull, it's a pity they hadn't done that before pushing me out, leaving our bags near the hedge I told Worra to keep close to me there was a dip in the ground and the cattle used to collect there to keep out of the wind, that's where I expect to find Old Bill our prize bull, setting off we scouted around the field which was quite large, arriving at the far side I became aware of a loud humming noise, it wasn't the space ship taking off this time holding Worra's shoulder I listened for the noise again, a glow appeared on the horizon then two beams of intensive bright light suddenly flashed across the sky, Its car headlights but what was a car doing travelling across the fields long after midnight.

Rustlers again the thought flashed through my mind if that's the case, we can let them find the bull for us then kidnap him by using our Tee guns, breaking into a run to get nearer the action I soon saw it was impossible, the vehicle was travelling too fast it passed us in the next field, and by the speed it was going there had to be a road there, years ago surveyors had put in pegs but we had driven the tractors over them, they never replaced them so we thought that was the end of the matter.

Now it looks like a new roads been built in the short time I've been away, crossing the next field to look at this new road I got a shock, it was a six lane highway, while sat on the wooden fence bordering the road two cars came thundering by in the reflected light I saw they were both sleek American looking models and I got the impression they were racing each other because of the speed they were going, standing on top of the fence to get a better view, I tried to find the farmhouse, I could see the roof but there were more houses clustered around it, then I suddenly saw a square of light come on where it should only be open fields.

Beginning to feel uneasy about the strange goings on I decided to get back to the ship and tell them it wasn't possible to get the bull, he was at the farm and it would be impossible to get him across the highway, Worra saw the sense in this and agreed to return to the ship and await further orders, only to find it wasn't there, the sudden appearance of the cars headlights must had frightened them away.

Feeling more than frightened myself I sat down by the tree where all my adventure had started, I felt really alone even though Worra was beside me, nothing seemed real any more even the tree pressing into my back didn't seem right standing up and stepping back to get a better look in the dim light I got a shock, the tree was massive much bigger than I remembered, frightening thoughts started to emerge but I brushed them aside not wanting to know what my brain was telling me, I didn't know what to do Worra was supposed to be in charge but she had no orders and didn't even know how to contact the ship.

In the end I had to give in and accept the inevitable, something had happened to the time scale everything looked unfamiliar because it was much older and therefore had grown bigger, many years must have passed for so much change to have taken place, Worra wouldn't believe me at first but when I pointed out the roofs of the houses and asked if she remembered the highway she had to admit it had changed, trying to work out what had happened only made it more confusing there if no way we can contact the spaceship, we are truly on our own from now on.

Finding out what the date is isn't as easy as it sounds you cant just ask a person what year it was without them getting suspicious, dawn was breaking through on the Eastern sky which told me the countryside would be waking up and we had to move on I heard dogs barking in the vicinity and got the impression there were houses near, our first job was to hide our kit which was easy, I remembered an old sheep dip I'd helped to dig on the edge of the wood, it was full of black water and dead leaves and stunk to high heaven the bags soon sank with the weight of gold in them and the guns we shoved inside a hollow log.

Making our way back to the highway we climbed the fence with the intention of thumbing a lift a succession of cars passed by at speed I gave each the hitch hikers thumbs up sign which they acknowledged with a blast on their horns, but no one stopped or even slowed, we had just cleared the outskirts of the woods when we heard a siren in the distance, looking back I saw a blue flashing light approaching at high speed, the last thing I wanted was to be arrested, grabbing Worra's hand I urged her to run clambering back over the fence we raced towards the cover of the woods.

The car screeched to a halt and I heard the two occupants shout for us to stop then they gave chase, we were nearly in the woods but I saw Worra was already tiring, and the heavy gravity was effecting me, "quick hide behind a tree" I told her "we can use our Tee guns to stop them, but don't let them see you," waiting till they got closer I let Worra fire first, her shot hit the man on the run he doubled up and ran straight into a tree, the second chap slowed to look at his companion when I shot him he just grunted tripped and fell on top of him.

Running over to them I searched their pockets for the car keys and money there was only a few notes in the wallets so I put a piece of gold in each one.

Little did I realise the trouble this was going to cause me later, they both had handcuffs hanging from their belts, so I handcuffed their hands around small trees. it would take them a while to climb them to get free.

They had devices on their belts which resembled the remote controllers we used for opening the doors and switching the lights in the spaceships, these I assumed were speech recorders to record evidence, to be on the safe side I took them with me to leave in the car, making sure the unconscious men could breathe properly we left them there.

= CHAPTER THIRTY NINE =

Starting the car was no problem and what a car, it had full luxury all over it must be a top of the range model, the engine was so powerful it ran away with me at first, but soon we were bowling merrily along, Worra was looking around in amazement, she had never been in a car before, she even told me to slow down as the speed frightened her, that struck me as funny she was a spaceship captain used to travelling at light speeds and was worried about going at seventy miles an hour.

She had found policemen's hats on the back seat and put one on, I put the other one on it may help as a disguise, she also discovered their lunch and two cans of beer in the glove compartment this was very welcome we hadn't eaten since the night before.

Suddenly the radio crackled into life, making Worra squeal in alarm, Calling car Four O Five Eight, code Seven One One report on your hitch hikers, she didn't understand what was being said and answered in her own language, which must have sounded Chinese to the operator on the other end, then came a sharp message loud and clear, to stop playing about and report directly to the superintendents office on arrival at headquarters.

Realising the police may have second thoughts about our funny message its obvious they will be looking for this car, seeing a sign for City Centre made my mind up, I automatically followed it we could dump it there without anyone noticing, the traffic was getting heavier and we would soon be at a stand still, flicking on the siren would only draw attention to ourselves, and upset Worra on the other hand it would get us through the traffic, otherwise we would be trapped, warning her I switched the siren on.

I pulled out of the line of traffic put my foot down, and sped on to the head of the blockage, a motor cycle cop was directing the traffic, on seeing our approach he waved us through, catching sight of his motor bike I nearly ran him over in shock, it was massive all chrome and glitter just like the Harley Davidson bikes the American police used in old gangster films, realising he could report seeing us and easily catch us in traffic the sooner we left the car the better, turning off into a side street I had to brake sharply when a woman ran into the road and banged on the windscreen.

"Thank God you're here quickly their killing him" she shouted, a small crowd attracted by the commotion had gathered and we had no option than follow her she stopped outside a small shop, and pointed inside, quick in there, and pushed me in the back so unexpectedly that I fell over the doorstep and fell sprawling on the floor, trying to get up I was kicked in the ribs and fell over again, then I heard a grunt and a heavy weight fell across my back pinning me to the floor, I had to struggle to get to my feet, now I took in the situation, an old chap was tied to a chair and had been badly beaten up by the look of the blood running down his face.

Worra following me in had seen two chaps set about me promptly shot them both with her Tee gun, then pandemonium erupted inside and outside the shop, the woman was having hysterics while wiping blood off the old man's face and trying to thank us at the same time, the crowd outside were trying to get in to see what was going on, this was the last thing I wanted to happen of all the streets to turn into I had to pick one where a robbery was taking place, the woman had phoned the Police and they could be here any minute.

This thought galvanized me into action, I snatched Worra's police hat off her head and stuck it on a young girl nearby, then picked my hat of the floor and stuck it on a gawky looking lad who thought it was hilarious and pranced around preening himself, this suited me fine and I took advantage, steering Worra out through the door and up the road, as we turned the corner a police car passed us, no doubt answering the woman's phone call call from the shop.

Walking along I was amazed at the number of people shopping and the amount of goods for sale, if I was amazed Worra was spellbound, I felt sorry for her, being hurled into a totally different environment, she had seen pictures in Toms books, but to see it in reality was a shock.

Although not as much as I got on checking the dates on the loose change I'd got off the policemen and found the latest date of two thousand and four on it, Bloody Hell we've jumped forward fifty years, Worra couldn't understand why I suddenly sat down on a shop window ledge to recover from the shock, when I explained she wasn't concerned she just shrugged her shoulders and said "we thought something like this would happen."

"We believe the founders do it to stop anyone believing anything you tell them, it happens on outward journeys and reverts when we return to our own level," dashing over to a parked car I bent and looked in the door mirror, the reflection was of a young lad of twenty, breathing a sigh of relief I shouted at Worra.

"You mean to tell me you knew it would happen," Yes" she said "but if you had known, you wouldn't have come," I was appalled at her indifference, and yet I had to admire her honesty.

Having a lot of questions to ask and needing time to absorb the information she had told me, I sat her down at a café table we were passing, a waiter came out immediately, which reminded me of a spider pouncing on his victim, Christ he was Chinese have they won the Korean war and taken over the country, the though was instantly dispelled when he spoke in English, then he turned to Worra and spoke to her in Chinese, she was too shocked to answer him and spoke to me instead, "This man speaks our language how can that be" the waiter understood and spoke to her again, and she answered in a lengthy conversation, the man then called out to others working in the shop, they came tumbling out with excited cries greeting her like a long lost relative.

With all the chattering going on my translator overloaded and switched itself off, I felt a little apprehensive not knowing what was being said I'm sure she wasn't telling them she was from another world, otherwise they would be laughing at her and not with her.

While all the talking was going on I checked out the menu, all the prices were shown and I soon worked out the value of the money we had twenty five pounds in notes and eighty pence in coins which was enough to buy a meal, it was getting late and I was wondering where we could spend the night, tapping Worra on the shoulder to get her attention I motioned for her to switch her translator on she immediately started to tell me about her new friends noticing they were looking at me In a funny way.

I realised they were puzzled by the way I could understand Worra but I didn't know what they were saying,. thinking fast I put my hand under Worra's chin and turned her face towards me and looked intently at her mouth as if I was lip reading, it must have looked authentic because they looked away as if they were embarrassed I told her to order a meal and we had to find somewhere to sleep, as the police would be looking for us, the waiter spoke to her, then spoke to the rest of the people around us they nodded their agreement, then he spoke to me in English.

You and your wife are welcome to stay here for a few nights if you would help in the café our son and his wife have returned home for a short holiday, we are interested in your wife's stories, she is well versed in our ancient history even her language is from our distant past, then he asked me where we had met, not knowing what Worra had told him I had to think fast "Oh I said that was when I saved her life but I wont dwell on that she don't like people to know."

Just then our meal arrived saving me from further questioning, I hadn't had a proper meal like this for ages and really enjoyed it, even finishing off with a bottle of Chinese wine, I insisted on paying for the meal but they wouldn't accept anything, this started to make me suspicious of their motives for wanting us to stay, with us being the only customers the head waiter said they were locking up early, they showed us the flat, then locked up and left us Worra was like a little schoolgirl loose in a toy shop running the taps and examining everything in the flat, eventually she tired of this and I showed her how to work the shower.

I lay on the bed thinking over the events of the day so far we had been dead lucky, thinking things over it seems too good to be true ending up in a place like this, not many people would take in complete strangers and give them the run of the place and not accept payment, That's it that's what's been troubling me at the back of my mind, anything that seems too good to be true, is too good to be true.

What are these people hoping to get out of us they know the police want us they know we haven't any money, it must be Worra they want, laughing at myself for being so melodramatic, I tried to clear the thoughts from my mind, but the more I tried to clear them the more they came back, in the end just to satisfy myself I wedged the chair back under the door handle checking around to make sure there were no other entrances I put our Tee guns handy just in case.

Worra couldn't understand why I was doing all these preparations, I tried to explain the situation but she wouldn't believe these nice people would do us any harm, to be on the safe side I decided to sleep on the floor in the end this proved to be a good idea, trying to keep myself awake was impossible after such a hectic day we were exhausted and soon fast asleep.

A small scratching noise awoke me and at first I thought it was a rat then it changed to a rasping noise, by the light of dawn just breaking, I saw the door handle turning, whoever was trying to open it cursed and I heard a lot of whispering going on, then suddenly crash the chair splintered and the door burst open, a body came hurtling across the room, followed by two more, one made for Worra and the other came for me, I took a snap shot and hit him in the face he buckled at the knees and collapsed in a heap turning I saw the other chap bending over the bed, I shot him in the back and he collapsed on top of Worra she squealed and wriggled out from under him, hearing a noise behind me the chap who had smashed the door down was coming at me with a knife, before I could bring my gun up he grunted and fell over, his knife thudding into the floor beside me.

Worra was sat up in bed looking very pleased with herself, I had to congratulate her on her shooting, most girls would have fainted or hidden under the bedclothes, at least I don't have to explain my actions to her she can see for herself they meant us harm, telling her they wanted her for the white slave market, wouldn't be a very good idea, in fact going by her performance in the cabin she may well have gone with them willingly.

= CHAPTER FORTY =

Surveying the smashed door and the three unconscious bodies I felt we had outworn our welcome and the best thing to do was to get out before the others came to open up, Worra didn't think this was a good idea, she said the dogs would be waiting outside, we shouldn't be going until it got light, I knew there were no animals out there but there was sense in what she said.

Our three assailants were still out cold but they could recover at any moment, so we blindfolded each one and tied them to the radiators, I don't think they would recognise us again, events had happened so fast they had been rendered insensible the moment they barged into; the room.

Seeing no sense in rushing out into the freezing cold at five o clock in the morning we stayed put and cooked ourselves a hearty breakfast, keeping our Key guns handy in case the owners came to start work early, having a last look around for anything that would come in handy I found two waterproof coats that were ideal for covering up our space clothes. Worra even remembered how to open the till and helped herself to the money in it before we let ourselves out of the front door, she had seen me emptying the pockets of the unconscious policemen so I'm afraid It was my fault for setting her a bad example.

Locking the door behind us, we set off along the street, it had just turned seven o'clock the few people about were on their way to work and were too preoccupied keeping warm to notice us, I was at a loss to know what to do none of the shops were open, and I can't remember where any transport café's where, a group of people were standing on the edge of the pavement and I thought they were waiting to cross the road, just then a bus pulled up and they all got on, catching Worra's arm I followed them, it would be warmer on the bus than wandering about in the cold.

The driver shouted "Ticket" at me and I shouted back "Its okay I'll pay the conductor," everybody burst out laughing, except the driver who held out his hand and mumbled "seventy five pence each," not being sure of the value of the money I held out a handful of coins and he helped himself, gave me two tickets and prepared to drive off.

Worra didn't know where to look first, I'd forgotten this was a new world to her, she had never seen houses, shops, cars or so many people before in her life, she was having more things to get used to in my world than I had in hers, the bus stopped a few times and I recognised the route, the terminus was at the Virgin Tavern, the place where I used to drink, this seemed a funny coincidence I've come a complete circle back to the place where I started from, although I doubt if I can get a drink at this time of day.

The sun had broken through the clouds by the time we got off Worra nearly fell over a few times trying to watch the drifting clouds, so in the end I stopped to let her look at them, waiting for her to lose interest, I looked across the road at a garage that had been built there with a few cars for sale, this interested me, a car would be ideal for getting about in, going over and looking at the prices soon convinced me we hadn't enough money with us to buy one, an early bird salesman was opening up and soon came over to enquire if I wanted any help in selecting one.

I said they were too dear for me and explained I was home on leave from the army and was due back in a weeks time, he said he had been demobbed last year and asked what unit I was in, of course I gave him the first unit that came into mind and said the Glider Pilot Regiment, he looked hard at me and said your kidding they were disbanded in the nineteen sixties, now I had to think fast, not all of them I replied quickly, a section was left operational to fly officers to meetings and ranges this seemed to satisfy him and we continued chatting for a while.

Then I dropped another clanger by asking him if the Knights still ran the pub across the road, I doubt it he said “they all died in the sixties, but if your interested why don’t you come and have a drink in there tonight, we’ve got a dart match on and you’d be welcome,” this would present a good opportunity to find out what had happened to me after I had disappeared, promising I would be there I was about to turn away when he said.

“Hang on a minute I’ve got a part exchange around the back that might interest you,” following him into his office he collected the keys and we went to look at the car, it looked the worst for wear but it started up first time.

“Tell you what” he said “Give us a tenner and its yours for a week,” well I certainly had enough money for that and handed it over straight away in case he changed his mind, the car was even taxed to the end of the month, I couldn’t believe my good luck in getting it so easily, then he spoilt it by asking for my address or army number, because I hesitated before I could make one up, he said never mind, if you get picked up by the police I’ll say you stole it and I hadn’t noticed it had gone.

Checking the car over to make sure there was oil and water in it I drove it around to the pump and asked for four gallon of petrol, the chap laughed, “where have you been for the last few years its all in litres now,” excusing myself by saying I had been in the far East for a while I called Worra over, seeing her oriental features confirmed what I’d said, following him into the office to pay I had a shock the petrol price was more than twice the price of the car.

Now I was at a loss what to do, driving around the city could be dangerous, the road signs were different and I’d seen cameras surveying the roads and pavements, the way the cars were crawling along I expect they were to stop speeding it looks like someone watches the screens and radios to the nearest police car to catch the culprits, having nothing better to do I drove around the district where I used to live, passing my old house I saw an old man entering the gate, stopping and reversing back I called out ”Excuse me do you know where the Clayton's have moved to, He swung round and said "they haven't, they still live here," if he hadn't come over and stuck his head in the car window to get a better look at me I would have driven off, instead this old chap goes white, swears, then shouts "Our Stan.

I had already thought what to say if I met anyone who knew me before I'd left Earth so I had an answer ready, "No" I said" "he was my dad, he used to tell me stories of where he used to live" "Come on in "he shouted all exited, Marie will want to see you, No" he said "I'd better go and tell her first, you're the spitting image of our Stan and she'll think you're a ghost," as he disappeared down the path it crossed my mind to drive off and save myself a lot of difficulty explaining.

Then I changed my mind I couldn't be that heartless, they must have been wondering all these years what had happened to me,, Besides I'd like to know what's been happening while I've been away, remembering I lived in a pretty rough district I locked the car then hearing footsteps approaching I quickly turned only to be fiercely grabbed and hugged by an old woman who had come running out of the house.

This had happened so fast I didn't have time to see who it was, assuming it was my brothers wife I didn't fight her off, I was more concerned about Worra shooting her, if she thought I was being attacked, however I didn't worry the tears streaming down the woman's face showed she meant no harm.

My brother rescued me by catching hold of my arm and inviting me in, we followed him through the front door and I was amazed at the alterations that had taken place, I'd expected to see the old Triplex fireplace with the shelf above it and the ovens where we reared chickens for our Christmas and New Year dinners.

= CHAPTER FORTY ONE =

Instead there was now a gas fire blazing away, with armchairs that looked so inviting I sat down in one and it felt like I'd never been away I felt so comfortable I started to close my eyes.

"Don't you look like your dad," the voice brought me fully awake We've always wondered what happened to your dad, now you're here you can tell us the real story," I certainly will" I replied, "we used to talk a lot about his family and neighbours, I feel I know them all personally, the Langfords, Patricks, O'neils, Masons, I could name all the people in the street, but first you tell me about my dads disappearance and I'll take over from there," I needed time to think up a story to fit in with theirs.

My brother interrupted at this point and continued, "when your dad disappeared we all thought he had been murdered pieces of human bones were found but his gun and belongings were never found, another chap who lived at the farm disappeared at the same time, he was quite a mystery man the police couldn't trace him, some rumours went around that he had killed your dad and pinched his belongings, we still get letters in the papers trying to solve the mystery, evidently your dad wasn't killed otherwise you wouldn't be here."

"Tell us what did happen all those years ago were dying to know" "OK" I replied "Dad told me that while he was waiting for some foxes to return to their den, he remembered he had a letter in his pocket which he had received some days earlier, he hadn't bothered to read it as he thought it it was from the inland revenue with a tax demand, on reading it he was alarmed to find It was an urgent recall to his unit.

He buried his gun and kit and hitch hiked to Middle Wallop, I remember the name as its quite funny, the Yanks used to call it centre punch when they used it in the war, anyway he ended up in Korea, that's where he met my mother she worked at the American army base there, dad spent some years as a prisoner in North Korea he was released but died soon after, I went to live in the states with mum and when she died I joined the army, and now I've served my time we decided to have a look at the old country., and a holiday at the same time.

While we were talking I noticed my brothers wife looking at me in a strange way, in the end she couldn't contain her curiosity any longer, and said.

"When you look towards the light a scar on your forehead shows up, Your dad had one in exactly the same place, he got his chopping firewood, a sharp piece flew up and stuck in his head he was lucky it missed his eye that's why I remember it so well, how did you get yours, surely it wasn't chopping wood, no one lights fires any more", I'd never thought of my scars giving me away and tried to make up an answer quickly, by saying I got mine fighting a girl, and added they were rough and tough at my school.

She didn't laugh and I could see she didn't believe me, because her next question threw me, she said I expect you have a big scar on your knee, your dad got his when he fell down the steps carrying a bundle of rags to exchange for day old chickens.

Thinking I was being clever I rolled up the wrong trouser leg, this didn't have the effect I expected, She squealed.

"I knew it you are George's brother Stan " telling her not to be so silly I appealed to my brother.

"How can she think that, she only has to look at me to see it's impossible,"" No" she spoke to George "the scars on that leg are ones he got falling off his bike on a cinder path, there are still two bits of cinder embedded in his leg which show up like birds eyes."

My brother was embarrassed and tried to talk her out of her accusation but she couldn't be swayed, in fact she became more insistent, "Al-right she almost shouted "get him to show his other leg if there is no big triangle shaped scar on that knee I will admit to being wrong," feeling like a condemned man I had no option, rolling up the other trouser leg produced another loud squeal, the big scar on my knee was a dead give away, "There I told you so," she shouted triumphantly, not realising her implications.

This was followed by dead silence as the enormity of her discovery sank in, George was the first to speak.

"Come on" he said, "Tell us we are all dreaming and none of this is true, you are supposed to be dead, we attended the funeral of the bits of bone that were found, so you owe it to us to tell the truth, if Marie saw through your story anyone else can," seeing the sense in what he said I put my brain into overdrive to come up with a story to cover all the situations, in the following tale I think I did brilliantly.

I explained that I had been working behind enemy lines in North Korea, and got captured, they used prisoners as guinea pigs to experiment on with age retarding drugs, they wanted their president to live for ever, over the years the other prisoners died off and I was the only one left, the guards began treating me with respect especially when I said I would kill myself and they would be in trouble, I even went on training runs with two guards and Worra my nurse, she had contacted an American peace mission whose ship was lying in the harbour I pretended to be my fathers son and they arranged my escape, the two guards were killed in a tunnel leading to the harbour, a car picked us up and we hid in the boot while the car was lifted aboard.

The Americans spirited us away to the states, and I was interrogated for the names of prisoners who had died and the methods used by the Koreans, I never mentioned about the age retarding programme otherwise they would have kept me under observation and never let me go.

"Wait a minute George interrupted if you spent all that time in Korea why don't you speak to your wife in Korean, That's easy I replied and used the bits of her language I knew to speak to her, if it sounded double Dutch to George it must have sounded worse to Worra, however she had been following our conversation closely and understood English more than I gave her credit for, she switched off her translator and spoke in her language this satisfied him and I carried on.

"The Americans couldn't understand why my dad had never mentioned me in his messages, I told them the only reason was it would have identified him to the enemy," then I joined the US army served my time and decided to visit the old country to see all the places I remembered, I ended up by telling him I couldn't say any more as a lot of what I've said already is top secret and they mustn't tell anyone.

The Koreans had tried to kidnap me in America and the CIA faked my death in a motor accident to throw them off the scent but they will still be looking for me, they don't give up very easily, the Americans were also wondering why the Koreans had been so interested in me and I got worried if they found out I was the results of the drug ageing experiments I would have been confined out of sight for the rest of my life, while they experimented on me, I finished up with the plea that now they can understand why I tried to keep the truth from them.

Finishing my story I sat back to watch the reaction, Marie was the first to speak, and being a typical mother after hearing my Earth shattering story she said "have you had any breakfast yet, Well I could have kissed her, she made me realise I was back in the land of normality, while she was cooking I asked my brother what had been happening while I've been away, I meant in the family, but he thought I meant in general, "Oh" he said "Everything is going wrong now the Nazis are running the country," what do you mean" I shouted "how the hell did that happen, we won the war and the Nazi leaders hung," "Yes they were," he said, "but they left money for German industrialists to buy up Western businesses, they used Nazi methods promising lucrative jobs in the new order to join them, they now make our laws and run the country.

I said "no government would be so stupid, and the people would never let it happen, Besides Princess Elizabeth had only just signed the Coronation Oath to stop outsiders ruling us, if she had broken it she would have been kicked out of Buckingham Palace". He admitted she still lived there, that convinced me he'd got it all wrong, I told him I wasn't interested in politics! had enough troubles of my own, he asked what my problems were and I told him about our mix up with the police on the motorway, and I'd left the States on a holiday visa and had no driving license, insurance, or identification papers.

He laughed and said that's no problem, all the old crooks and their families still live around here, they'd skin a gnat for halfpenny, I also asked about papers for Worra, he reckoned it would be better and cheaper if I got married legally at a registry office, then Worra couldn't be deported as an illegal alien, I thought that was quite funny considering she would be the first genuine legal alien here.

Now I mentioned I'd have to pay in gold or get it changed into cash, I thought this would be a big problem but he just laughed and said it couldn't be better, as most of the crooks had moved up market, and would do anything to get their hands on gold, this suited me fine I had been wondering how to make contact with a dealer, now all I had to do was collect my gold from the woods where we had hidden it and change it for cash.

Thinking about cash reminded me I hadn't got a bank account, and I couldn't get one without identification, Marie came to my rescue, she was Irish and told me the IRA terrorists get a copy of a dead persons birth certificate from London, and use it to get a British passport, I wasn't convinced it would be that easy but she insisted it was so well known in Ireland expectant teenagers use the same method to apply for a British passport to come to England to have their babies and a free All expenses paid holiday, on the British national health Service.

= CHAPTER FORTY TWO =

My brother and his wife really entered into the spirit of adventure, by helping me, I suppose it made a change to an otherwise drab life, first they insisted I stay at their house then they set about disguising us, me with a haircut and Worra with make up, it was comical to see Worra trying to put on lipstick, then George insisted we used his car in case the police were looking for mine, leaving Worra we set off to collect my belongings, going in a roundabout way to the woods I was amazed at the number of new houses that had been built, driving past the old farm I saw it was completely surrounded by new houses.

Even the cowsheds and barns had been converted into homes, remembering the years I had spent clearing cow muck out of them, and now people were living in them, I think Cecil the old farmer would have laughed to have seen this, my brother told me he had died in nineteen eighty seven, and buried in the local churchyard.

We had a bit of trouble getting to where my stuff was buried, it wasn't possible to leave the car on the motorway which was the closest, and to be seen carrying a gun would result in the whole police force turning up armed to the teeth, eventually by opening a few gates and cutting across the fields we arrived at the hiding place, Setting Arthur up as a lookout I scraped away the leaves and soon found the gold and packets of stones, when I went to collect the musket I had the shock of my life.

I had pushed the gun down a big hollow log out of sight then rammed in some big stones to block it, now the hole was open and the stones lay all over the place and there was no sign of the musket, looking at the devastation reminded me of the times I had seen the same signs, Foxes had done this, lying down and reaching into the hole I felt the butt of the gun wriggling it around I managed to drag it out, what a relief to find the diamonds still in place, pulling the plug of rag out of the gun barrel resulted in a cascade of pea sized granules falling into my eager hands, collecting the kit together we put it in the car, just in time, a gang of teenagers came by and wanted t know what we were doing, We told them we were laying snares to catch rabbits and if they didn't clear off and leave our snares alone they would be in trouble, saying this ensured they would obliterate our tracks looking for non existent snares.

Arriving back at the house, I marvelled at the transformation of Worra it was amazing the difference make up and hair styling did for her appearance now she looks quite attractive. and is revelling in the change, she keeps looking in the mirror and flicking her hair into new positions I hate to think what will happen when, we return to her home this started me thinking about our position here, we have no papers and very little money, I've got plenty of gold and uncut diamonds, but I cant just walk into a jewellers and change it into cash, the best thing I can do is contact the crooks George told me about, find out if they can change it for me.

George arranged a meeting for that night, at the Tavern this suited me fine as I had promised to attend the darts match there, it was nothing like I expected I was introduced as Stan's lad who had been living in the States and returned when it got too hot for me.

This explained why I had the gold and no papers, I said Worra was one of our gang who worked at the bank we collected the gold from, I had decided on the way to pretend I had a tough American gang behind me who didn't stand for any monkey business, and were looking to unload more on a regular basis, so we would need bank accounts to transfer the money into.

Our team lost the darts match which was a bad sign, I had been accepted into the crooks circle on my brothers recommendation and the fact I was Stan's son, but it didn't seem right to me they hadn't asked any questions like how did I get the gold into the country, I had already thought of a way which would have gained their confidence, by telling them we brought one lot over by coating gold balls with lead making them look old and passing them off as musket balls from the wars of Independence' but one of the shipments accidentally went to the British Museum, so we had to find another way.

I told them we had found one and now my bosses want small units all over England to convert gold into cash, they argued at first about the exchange rate I offered, but I silenced them by saying it wasn't negotiable they could either take it or leave it, after conferring they agreed, now they showed their professionalism, the weediest looking chap in the gang produced the equipment to weigh and analyse the gold balls the Homeworlders had given me, I held my breath as he did the acid test, I hadn't thought of doing that, at last he nodded in satisfaction, we agrees on the price and they paid in cash, I asked for a receipt which my bosses wanted signed by all of them.

Once the dealings were over the gang gathered around me and wanted to know what had happened to my dad, now I wondered if I had been wise to pick this place for a meeting, but my fears were dispelled when I said I cant tell you walls have ears this produced howls of laughter I didn't think that was very funny and said so, then I was told this place was their headquarters and no one lived long who squealed on them, so I told them the first story I had told my brother, which seemed to disappoint them, I think they were expecting an account of gory murder in lurid detail.

Now I had plenty of money I sorted out the chap who had sold me the car, and did a deal to change the old banger for a newer model, as I had been warned the police stopped and checked all old cars, I suppose I had been lucky so far, which made me realise I was still in a dangerous position surrounded by this gang of hardened criminals I wasn't sure they would just let me walk out with such a large amount of money on me, so I paid cash for a car I hadn't seen, and gave six months rent in advance to my brother, this appeared to relieve some of the tension in the room especially when I said I can put this outlay down to expenses as my bosses gave expensive funerals, I could see the murmurs' of Mafia surge around the room, which certainly made me feel safer.

The car dealer opened up the garage across the road and selected a nice car for me, I think the threat of the Mafia gave me a better car than I would have got otherwise, I filled in a form giving my brothers address and he said he would complete the rest of the details, with all the necessary papers to make me legal, feeling I had done a good nights work I went back to the pub and bought everyone a drink.

It was funny waking to the sound of rain on the windows, Worra had seen rain the day before but she was still mesmerised by the way it spattered on the ground, I wondered what she will think when it snows, although winters a way off and its quite warm so our weather doesn't seem to affect her as much as I expected, her world had a constant warm temperature, but I noticed she was still wearing the underclothes from her world and they had very good insulation properties.

After a couple of weeks of driving about the country sight seeing, lazing about and taking in the sights even Worra appeared bored she couldn't understand why everyone rushed about, and spent so much time in shops collecting possessions, she said why get them just to lock it away to stop others taking it off you, the smell of car fumes made her feel sick, and I think she missed the obligation system I had to admit she had a point, I missed the fliers the fresh air and the excitement of all the new things on her planet, now I jumped every time I saw a policeman, and there was always a gang member tailing me, I expect they were trying to find my contacts for the gold.

This constant looking over my shoulder was beginning to worry me when I told Worra I would like to return to her word she certainly brightened up at my suggestion, but she reminded me of the time difference, and the fact we had no means of contacting them, I said the spaceship that brought Rip Van Winkle back to Earth returned to her planet in normal time and her people would know about the time lapse and make allowances for us, this satisfied her on that score but she said we still have no way of signalling them.

I had an idea that we could use our translator signals beamed into space by a powerful transmitter and hope they would be picked up, but Worra dropped a bomb shell on that idea when she said they didn't use radio waves on her world because they interfered with the electrical power transmissions, I pointed out the. auto brains communicated with each other, but she didn't know how that worked, now we were back to square one.

Trying to think things out rationally I kept asking her questions about how they contacted previous agents on Earth, when I asked about the space beacons built into the Pyramids, she became very evasive, I knew there was no way we could go to them because of passport controls, our papers would soon be found as forgeries.

The more I pressed her the more she became agitated, working on the assumption people only speak the truth when they are mad, I really went all out to upset her by saying she had been telling me lies about her space travels her ancestors and anything else I could think of, then the truth emerged, there had been no intention of returning me to her world.

Once the bull had been found and loaded aboard the spaceship, her orders were to shoot me unconscious, change my clothes, remove any thing from her world and leave me lying in the field, I asked her how she would remove my translator as I understood they couldn't be taken off she just shrugged her shoulders and said I would have cut your head off, I'm sure she wasn't joking, so it was only the sudden appearance of the car on the Motorway that saved my life.

Now I understood a lot of the things that had been bothering me, like why was I allowed to collect items and gold to bring back or why we had no orders or contacts to meet, I even doubted the existence of the Founders, I had seen three dimension holograms here on Earth which must be how it was staged.

Worra was in quite a state when I finished my interrogation, but I thought it was worth it, now I know what I'm up against, first I calmed Worra down and assured her I would help her to return to her own world on condition she took me with her, this she readily agreed to, even when I said she could be under sentence for not carrying out her orders, she just shrugged and said the duchess would help her when the time came.

This puzzled me until she explained, when a person of high order is sentenced for disobeying orders, which normally carries a death sentence they can appeal providing a higher order person assists them and the original sentence can be adjusted according to the severity of the results of disobeying the order.

It's funny how my life has suddenly reversed, before I couldn't wait to get back to Earth and now I cant wait to leave it, which could be quite some time as clear skies don't come very often, in the meantime to keep myself occupied, I bought a computer, and attended an adult class, who all laughed when I tried to give it instruction by talking to it, these computers are good but nothing like the Autobrains, they can think for themselves.

The next full moon found us all packed ready to go but it was overcast and drizzling with rain, which was a disappointment but it couldn't be helped; to pass away the time I had started to write a story about my adventures on a parallel world, it will certainly be classed as fiction so I can put everything in it, with no fear of anyone believing it, when we eventually leave I shall leave all the gold with my brother to pay off the gangsters at the pub.

The frosty nights are coming in now which means we can expect some clear moonlight skies and the time for a full moon isn't far off, Worra is getting exited anticipating our leaving, I'm rushing to get my book finished and now a complication has cropped up.

Arriving home from a shopping trip I found a strange car parked outside the house, of course living in a tough area I immediately thought of burglars and put plan A into action. Knocking on the front door it was opened by a tall chap with another one standing behind him, holding up the box of grocery's I said your groceries Mr Clayton just sign the receipt and I'll bring the rest in.

I handed the box to the tall chap, he in turn handed it to the chap behind him, then he put his hand in his pocket and produced a police warrant card which he waved under my nose, I had intended to collect my Tee gun from the car and use it, but if these policemen were genuine I wanted to know why they wanted me.

Stalling for time, I offered them some grapes off the top of the box, and told them I would cross them off the list, this put me in their good books and they soon opened up, telling me they only wanted to question the chap who lived here, about rumours concerning a chap who had disappeared fifty years ago.

I asked to see both their warrant cards then explained who I was, and told them I thought they were burglars, after they asked me a few questions and I told them the missing man was my father, but I couldn't help them as I had spent most of my life in America, I called Worra in from the car and she made tea for us, in the course of our conversation I mentioned I was writing a book little realizing the enormous impact this would have on our plans to return to Worra's world.

She thought they were quite friendly but I was glad when they had gone, they were sure to check on my story and I know my forged papers wont stand up to any investigation.

Things are beginning to come to a head, I've finished my book and sent it off to an agent to be checked and published, this will take months and I shall be a long way from here by then, I only hope, I'm not here when its on sale, the police will certainly be interested in reading it, if its only to clear up a few unsolved crimes like who tied up two Motorway policemen and stole their car.

No one will believe my story about stowing away on a spaceship to another planet but I'm sure someone will think its feasible when they put all the clues together and there must be plenty of those, they will wonder where did I get the gold and uncut diamonds, my advanced knowledge and how did our Tee guns shoot such a distance.

I felt a moment of alarm I had forgotten about them, if the police search the house and find them I'm in deep trouble, I can't get rid of them, its only one day away from a full moon and we may need them to stop being arrested before we can get away I'm beginning to think we wont make it, the police are asking a lot more questions, the gang I sell the gold to are still following me.

And I'm concerned about Worra she is certain the spaceship will appear at the pick up point, but I get the feeling she is keeping something from me, I even wonder if she's in contact with the ship and had only recently got the orders to stop me returning with her, she had told me the translators didn't use radio waves as it interfered with the power transmissions but I don't think she is telling the truth I remember the Duchess using her translator to contact Worra to bring some footballs with her.

The more I thought about being double crossed the more likely it seemed possible, she will need my help to get to the landing site, so if she is planning to stop me it will happen then, I weighed up the problems, I've got the Keystone cops, the Crazy gang and I suspect Worra all after me, the obvious thing is to get them to cancel each other out, and I have an idea how to do it the diamonds I'd brought were only poor quality but valuable enough for use in engineering and would still fetch a good price, these will have to be sacrificed as bait to lure the gang away from tailing me, and it would only take an anonymous call for the police to divert their attention, to the gang.

= CHAPTER FORTY THREE =

This leaves me free to concentrate on forestalling any attempt by Worra to leave me behind, trying to think of all the possibilities only gave me a headache, I've already decided she needs me to get her to the pick up point, so I wont worry about it till the time comes.

The fateful day arrived with a thick coating of frost which blanketed the hedgerows and kept the temperature below freezing all day, I had kitted myself out with full winter clothing, but Worra didn't seem to feel the cold, which I put down to her wearing the highly insulated underwear she had worn on her own planet, when I asked her about it, she said it wasn't the clothes keeping he warm it was the effects of the radiation tablets she had been taking.

Bloody hell this was a bit of good luck, I could leave some with my brother to give to the scientists here, with a note to say I got them from North Korea, of course all my hopes were dashed when I asked her for some, only to be told she had given them to the puppy dog next door, angrily I shouted at her why had she done such a stupid thing, "because he was shivering" she said, well there wasn't much I could say after that, so I just said I was sorry for shouting at her, I know if I asked her why she hadn't given me any before she would have said because I hadn't asked, I keep forgetting she's an Alien and has different ways to ours.

As the day drew to a close I kept running the plans through my mind making sure I hadn't missed anything, my going away party at the Tavern had been organised and will be well attended as I promised to share out the diamonds, which I said were poor quality and not worth taking through customs, the gang had seen the police interest in me and knew I would want to dump them before being arrested.

So far I have the police trailing me the gang watching my every move and I'm not sure about Worra, although I know she needs me to get her to the pick up point so she wont do anything before then. this leaves the other two I have to distract somehow and the solution came in rather a bizarre fashion.

Phoning my brother from a call box I felt a slight breeze as the door opened a bit, I knew straight away it was my police tail listening in, so I pretended to fix up a gang meeting at the Tavern for a share out of industrial diamonds at eleven o'clock, I knew he had taken the bait when the door silently closed stopping the draught.

With luck my plan would see the gang and the police too busy tangling with each other to be bothering about me, and I'd be high tailing it to the pick up point with only one problem to solve, and that's Worra's teegun I can't disable it we may need it to get away, to be on the safe side I lined my jacket with silver foil and connected it with wire to metal strips on my shoes, I couldn't test it but I'm pretty sure it will work.

Half past ten found me in the middle of the most hectic party I had ever been to, I'd left two hundred pounds behind the bar for drinks and of course it was flowing like water, most were drunk or close to it, the bag of diamonds after being checked lay in the centre of the big table with instructions to be shared out after I'd gone, I said I had a plane to catch and it would take me too long to share out.

Saying to no one in particular I was going to the toilet, just in case anyone followed me, I went outside and looked at the stars with relief, it was a lovely clear frosty night just right for our ship to land or rather materialise. Worra was sitting in the car on the car park, she wouldn't come in to the party, she couldn't see any sense in mingling with a crowd and drinking till you fell over, she had the radio on and jumped up when I tapped on the window, she opened the door and I asked if everything was going to plan. She nodded, although I think she was too interested in the music to have noticed anything unusual.

Sat in the car I began to feel alarmed the noise from the party must have attracted the attention of the local residents, if they phoned the police it could wreck my plans I had allowed half an hour to get to the meeting place if they raided the place too soon, it would give the police time to trail me, and I don't want cars screaming up and down the motorway it will frighten off the spaceship like it did when we arrived.

The instant I switched the lights on, I found the car surrounded by figures in combat gear, both doors were ripped open and we were dragged out and pinned to the ground, a gruff voice said "bring her round here and cuff them together," now I knew it was the police and not members of the gang, who had grabbed us, I felt better, the gang would have tied us up making escape impossible and I'm sure I can talk my way out of being arrested.

I demanded to see the chap in charge who turned out to be the one I had given grapes to, he told me I had been arrested for questioning as the Americans had no records of me, but I would be in court tomorrow with the rest of the gang for disorderly conduct, then he shoved us in a police car telling the driver to step on it and hurry back with more handcuffs.

This was a set back to my carefully laid out plans but I wasn't worried we hadn't been searched which was a bit odd, but I suppose they were in a hurry to net as many of the gang as possible, I still had my tee gun which meant I could escape whenever I wanted to, which soon presented itself when the car had to stop at traffic lights. Checking to make sure no one was around I managed to get the gun out of my pocket, lined it up on the drivers hand on the steering wheel, and pressed the trigger, he looked at the red spots, grunted and fell sideways.

Moving into the front seats with Worra chained to me was a struggle, she pushed the unconscious man into the passenger seat then sat on his lap, the lights had changed so I turned around and drove back towards our pick up point, suddenly the radio came on telling all cars to converge on this area and stop a stolen police car with a man and woman in it, realising it would be madness to keep using it, I decided to call at the pub and get our own car back.

Worra still had the spare keys so knowing I had no choice I headed for the Tavern thinking about keys I asked her to look in the drivers pockets for the keys to the handcuffs, but when she started to look he began to recover, not having any compassion she just opened the door and pushed him out, I was appalled at this but couldn't do anything about it, so I just carried on.

We arrived at the pub in the middle of a pitched battle, the gang wasn't giving up their diamonds without a fight, the few police in the car park were too busy to notice us get into our own car which was still parked in the same place, I quietly drove out and once over the hill put my foot down, to get away from the place.

Speeding along and thinking how easy it had been to escape I wondered if the Founders had engineered it, dismissing this as a fantasy, I still thought it was funny that we had managed to bring all our sports equipment with us, turning to talk to Worra I saw she was holding her translator and talking into it, this confirmed my earlier suspicions that she had been in contact with the ship all the time, at first I felt angry at being deceived, then I thought perhaps it had been for the better, things were working out fine even being handcuffed to Worra is an advantage it means the ship will have to take me back with her.

A blue light flashing in the distance in front of us brought me back to reality as a Police car swept past at high speed, then came a jumble of headlights in my rear view mirror, I slowed down thinking he had crashed but next minute I was blinded by powerful headlamps right behind me, evidently he had done a high speed turn and was now sitting on our tail. I panicked to be switched from a feeling of euphoria to one of despair was mind numbing there was no chance of outrunning him, and we were only a few miles from our pick up point.

Worra tried to use her tee gun, I saw the red spots playing on their windscreen but it had no effect on the occupants, they were so close I saw the passenger using his phone no doubt calling his mates for backup, there was only one way to disable the Police car, I had seen numerous chases on television and knew the best way to do it, dabbing the brakes to let him know I was stopping, I slowed down until he was right behind me then snapped the gears into reverse, and gunned the engine.

The crash was much harder than I expected Worra squealed the back windows exploded in a great shower of glass and large cloud of steam gushed out, for a heart stopping moment I thought we had set on fire, selecting first gear I was relieved to find ourselves speeding away from the scene in no time, one of the headlamps had gone out which was going to make it difficult to find the gates when driving across the fields.

In my rear view mirror I had seen one Policeman standing by the broken down car and the other standing on the bonnet, this seemed peculiar until I realised he was reporting my position to other cars speeding to the scene, switching off my lights, made it easier it cut out the back glare and made it harder for anyone to see us, stopping at the first gate to the woods I was dismayed to see a large padlock securing it, turning the car to get a fast run to force a way through the hedge; I saw two beams of light bouncing up and down where we had left the road.

This meant a Police car was on our trail, selecting a place where it looked thinner I drove around to get up speed then aimed for it, CRASH, I burst through at speed a horrible scream curdled my blood until I saw it was caused by a single strand of barbed wire being drawn through the front wing cutting its way in deeper then it broke and we were free.

Now I was in home territory and put my foot down smashing thoughts gates and hedges in a wild Viking charge, trying to lose our pursuers, once over the next rise in the ground I thought we would be safe we can dump our gear at the pick up point, set the car on fire and send it back over the hill they are sure to follow it this will keep us out of view and give us time to get aboard the space ship, skidding to a stop I pulled Worra out after me, then swung her onto my shoulders she wasn't very heavy and it was easier to carry her than drag her, the lights of the Police cars were still a long way off which gave us time to unload and make the car ready for a diversion.

I had already worked out a plan how to do it and now put it into operation, ripping electrical wiring out of the boot I rolled the front carpet up tied it, and rammed it through the steering wheel so it wouldn't turn, next I opened the bonnet and tied the throttle linkage to the windscreen wiper arm tying knots in the wire until it would keep the engine at half and full revs.

Checking to see where the police were I found it was time to send it off, starting the engine and switching on the wipers I found it worked a treat, waiting till it slowed down I leaned in and rammed it into second gear, the wheels found more grip than I expected the car took off shoving me and Worra into a sprawling heap, luckily without injury. I intended to set it on fire but all the fag lighters were packed away with the sports gear.

While Worra was contacting the space ship I watched the car disappearing over the rise in the ground doing crazy zigzags as the engine speeded up and down, spinning the wheels it appeared to be doing a waltz, then the flashers came on and I thought the Founders had something to do with our escape after all, or did the rolled carpet move and operate the indicator switch, now I heard the noise of engines very close and people shouting they could see our car, the next few minutes would see if I was spending the rest of my life in jail or back on Worras planet living the life of Riley.

I felt the hair on the back of my neck stand up and thought it was Worra playing about until she said its here, hearing the whine of the ramp dropping, I turned to see our salvaged space ship glistening in the pale moon light, it looked massive compared to the original ship, unwrapping her arms from around my neck to give her the hint to get down she took no notice, in the end I had to shake her off, no one came to help us, so we had to frantically throw our luggage on to the ramp, then jumped on after it.

We were only just in time, the ramp had started to rise while we were still on it, and we both ended up in a sprawling heap on the floor, I heard a shout from the lift which had just stopped, and an answering shriek from Worra, then a red beam flashed across my line of vision and I saw the lift occupant who was aiming a gun in my direction crumple up and fall on the floor.

Worra had fallen on top of me restricting my movements, getting untangled I realised she had shot the newcomer, this I couldn't understand, just when I thought I was home and dry things had changed again, turning to her for an explanation I was surprised to see tears streaming down her face and she was pointing her tee gun at me, now she started to sound hysterical by shouting he was going to kill you and I've shot him, I have orders to shoot you and leave you behind, I only had to lift my hand to show we were handcuffed together which made that impossible, to be on the safe side I took the gun off her, then walked over to the unconscious chap by the lift, I saw he had a uniform on and a killer gun lay beside him.

I turned to Worra for an explanation which she was only too glad to give me, she said they had orders from the high council not to take me back, and if necessary to kill me, the penalty for disobeying the council was an automatic trip to the crushers, this turn of events puzzled me, the founders had wanted me to return, it appears they don't know what's going on, I put this to Worra and she thought it was highly probable because the high council had kept the drug situation from the Founders for hundreds of years.

Tying up the unconscious chap we dumped him behind our pile of luggage, then decided to try and release our handcuffs before doing anything else this was easier said than done, in the end Worra managed to slip hers off, this was a great relief it gave me freedom of movement, now I thought we could tackle the rest of the crew, we had two tee guns, a killer gun, and surprise on our side.

I thought about sending Worra to check out the top deck then changed my mind, the crew would wonder where the escort was, now there was no going back, the lift clunked to a stop and the doors opened, three crew were clustered around the control desk watching the black timer, it was still at the bottom of the screen.

This surprised me as it meant we were still visible to outsiders the three of them turned as we entered. I recognised the Duchess just in time, she shouted out a warning and I changed my aim to the second man while Worra dealt with the other one, not being sure if the crewmen were friendly we tied them up to make sure.

The Duchess ran over to Worra and hugged her, which was totally out of character, she wouldn't normally let anyone touch her, when I asked her why all the fuss she told me the whole story.

The guards had orders to kill Worra if she brought me aboard. this seemed silly to me why bring a ship all this way to pick up two people then kill them when it arrived. the Duchess said she'd explain more once we left Earth, then she told Worra to operate the controls. which started the black bar rising, with the emergency over she explained.

The High Council had decided we were a threat and wanted to get rid of us within their law, and it would be too late for the founders to do anything about it, Worra had been told to leave me on Earth, if she didn't obey the orders she would go in the crushers with me, the Duchess had been included in the crew as a witness for the inquiry the Founders would hold regarding my death, and the reason for Worra disobeying orders.

This was good news in a way, it shows the Founders are still a force the council take into account when dealing with me, I have a suspicion there will be a few empty seats on the High Council after any inquiry.

Now I won't have to watch my back all the time, I can find out about power generating, ask Merlin a thousand and one questions, see the other side of the planet find other factories visit other worlds especially the Homeworld.

I'm looking forward to exploring their world, if only half the things Worra and the Duchess have told me about it is true, there's certainly going to be many interesting things to report on, the possibilities are endless.

THE END OF A BEGINNING.

WS - #0079 - 290923 - C0 - 210/148/19 - PB - 9781780353470 - Gloss Lamination